The XYZ of CNC: Creating a Tool Path

By Logan Lambert

Prerequisites

There are several basic math skills that are required for you to have to get the most out of this book such as being able to plot a point on a graph, a basic understanding of fractions, and a basic understanding of decimals.

I have included a couple of reference guides in the back of the book and I will go into further details on a few of these subjects later in the book. But in addition to those a basic understanding of adding, subtracting, multiplying, dividing, will also be necessary for you to get the full experience of this book.

It is also a good idea to have some experience with basic machining. This book focuses mostly on the programming aspects of the CNC machine. A understanding how to calculate feeds and speeds would give you a better understanding of the process but is not covered in this book.

Acknowledgments

I would like to thank my family for their love and support throughout the years and a special thanks to my grandmother Elaine Pyle who went through this book and help check my spelling and grammar. And my baby sister Cypress whose constant want to help me with this book motivated me to actually finish it.

I'd also like to thank Stevie Lovelace and Mike Snowden, two great teachers from Northeast Mississippi Community College who taught me a lot.

I'd like to also thank my fellow coworkers who helped further my knowledge on this subject and a special thanks to Steven Ashby for checking my math on this book.

Table of Contents

Introduction

Welcome to the world of CNC programming. In this book I will be covering the fundamentals of programming basic movements for a CNC mill. I will go over the basics of the Cartesian coordinate system and how it's used in CNC programming. I will be covering the differences between absolute and incremental programming, going over G code and M code, and how to actually put together a program. Of course there's not really any point in going into this stuff if you don't know what I'm talking about, so first I like to go over some terminology. So let's get to it.

Basic Terminology

Basic Term

Absolute Programming- The style of programming which all points are based off of the part's home position. (Activated by G90).

Arc- A line that forms a radius.

Axis- A line of reference for the coordinates system, in CNC programming the X, Y, and Z axes extend out at right angles from a single origin point.

Breaking the chip- The process of releasing the downward pressure of a drill and partially backing it out of the hole to prevent a ribbon of metal from forming around the tool, or to help remove chips from the hole.

Cartesian coordinate system- A system that uses axes to plot points on a graph.

Chamfer- The operation designed to leave an angle in a drill hole or used to break the sharp edge of the work piece.

CNC- A abbreviation for computer numerical control.

Coordinate- A point on a graph.

Countersink- An operation designed to create a chamfer at the top of a drilled hole to allow a screw to set flush with the surface of the part.

Cutter compensation- A code that offsets the tool by a certain amount (G41 activates left compensation and G42 activates right compensation).

D code- a code used to specify a tool diameter when using cutter compensation. (G41 or G42)

Diameter- The measurement of a circle from one side of it to the other.

Drill- A tool used to make holes.

Dwell- A time delay usually used to allow a machine function to catch up with the program such as letting the coolant turn on.(Can be activated in a program by using G4).

End-of-Block- The end of a line of programming. To signify the end of the block placed a semi-colon at the end of line. (EOB is also used to abbreviate end of block).

Engraving- An operation that cuts a symbol or pattern into a part.

F code- A code used to specify the feed rate.

Feed- Term given to identify the movement of the tool into the part.

Feedrate- The amount the tool moves into the part in inches per minute (IPM).

Finish cut- The final cut of a cutting tool which usually takes off less material at a slower pace than a rough-cut to get a better finish on the part.

Fixture- A device used to hold the work piece in place while it is being machined.

Fraction- A number that is located between two whole numbers. (Part of a whole number)

G code- A set of codes based mostly around the movement of the machine activating functions such as rapid traverse, setting machine movement, and activating certain functions of the machine such as dwell.

H code- A code used to specify a tool height offset. (Activated by a G43).

Home Position- A fixed starting point for X,Y, and Z which all movements are calculated from. There are three sets of home positions relative to a machine. The first is machine home zero which is the position at which the machine returns to for a tool change. The second is fixture home zero which is the distance between machine home zero and the fixture. The last is part home zero which is the distance between the part and machine home zero or can be offset from fixture home zero.

I code- The code that represents the X axis incrementally from the starting point of a radius to the center of a radius during a radial movement (G02 or G03).

Inch- A system of measurement used in the United States.

Inches Per Minute (IPM)- Use to identify a feed rate in time.

Incremental programming- A style of programming which the origin point is redefined at each point (activated with a G91).

J code- The code that represents the Y axis incrementally from the starting point of a radius to the center of the radius during a radial movement (G02 or G03).

M code- Codes used for miscellaneous functions such as turning the spindle on and off, and turning coolant on and off.

Magazine- The placed in the machine where the tools are stored.

Mill- A cutting tool used to take material off the surface of a part.

Modal code- A set of codes that stay active throughout the program and require another code to deactivate them.

N code- Codes used to identify blocks or lines of data.

Nonmodal codes- A set of codes that only stay active for one line of programming.

O code- Codes used to identify the program number.

Offset- Compensation for a tool wear or a term given when setting its length.

Origin- A point of reference for all coordinates. The 0,0,0 point of the coordinate system.

Optional stop- A code that is used for stopping the machine. But will only activate if the optional stop button is pressed. If the optional stop button is not pressed the machine will ignore the code (Activated with M1).

Part- An object that is going to be machined.

Pick Drilling- A operation of drilling used to drill deep holes (Activated with G83).

Program- A set of instructions for the machine to follow.

Program stop- A code that is used to stop the machine. The machine will stop every time it reads this code regardless of any buttons active.(Activate with M00).

Q code- The amount the drill will travel before backing off to break the chip during a peck drilling canned cycle (G83)

Quadrant- One of the four sections in the Cartesian coordinate system.

R Code- The radius value for (G02 or G03) and represents the return point for canned cycles (G81-G89)

Radius- A measurement of a circle from the center point to the outside edge.

Rapid traverse- The fastest rate at which the machine can travel (activated with a G0).

Rough-cut-A machining operation that is used to get the most material off the work piece at a faster pace before the slower finishing cut.

S code- The code used to specify spindle speed.

Spindle- The component of a machine that rotates during the machining process.

Spot drill- A tool to use to locate holes for drilling.

T code- A code used to specify a tool number.

X axis- The line that goes left and right from the origin point on the coordinate system.

Y axis- The line that goes up and down from the origin point on the coordinate system.

Z-axis- The line that the spindle follows backwards and forwards from the origin point on the coordinates system.

Basic math skills

Math Skill

Like I stated earlier, having an understanding of basic math operations will greatly help you in the understanding of CNC programming. So for this reason, I will briefly touch on a few of the basics skills that you will need.

Fractions and decimals

Understanding how to turn a fraction into a decimal is a good place to start since the Cartesian coordinate system uses fractions in a decimal format. I have included a fraction to decimal conversion guide in appendix b, but in case you are unable to reference at any given point I will give you an example on how to turn a fraction into a decimal.

Let's take the fraction 1/2 and turn it into a decimal by taking the numerator 1 and dividing it by the denominator 2.

$$2 \div 1 = \qquad 2 \div 1.0 = . \qquad 2 \div 1.0 = .5$$
$$\underline{-10}$$
$$0$$

So the answer to 1 divide by 2 it is .5

What is the Cartesian Coordinate System

The Cartesian coordinate system uses points on a graph to plot the movements of a tool on the part. It is the bases for all axis movements. The point where all axes intersect is called the origin or X0, Y0, Z0. Now let's take a look at the basic form. Take a look at figure below and take note at negative and positive signs of the axes.

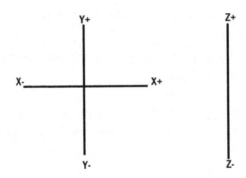

If you'll notice the graph itself is broken up into four quadrants each quadrant is assigned its own plus and minus signs for the X and Y axis

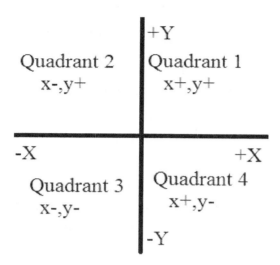

Now if you look at the figure below you'll see that each quadrant is broken up into several different smaller parts. Each mark (in this case) will be equal to 1 inch.

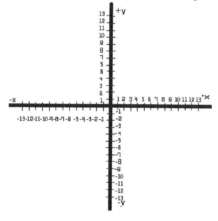

The origin point in the center (0, 0, 0) will be represented by a certain point on the part. Then using a true to scaled measurement a point will be placed on the coordinates which you wish the tool to go.

Using the Cartesian Coordinate System

Now you are familiar with the design of the graph I'll now show you how to use it. First let's look at a pair of coordinates.

-2,3,1

Let's start off with what each number represents. The -2 represents the X coordinate, the 3 represents the Y coordinate, the 1 represents the Z coordinate. If you forget what number represents which coordinate just remember that it goes in alphabetical order, which means the X is going to be on the left, the Y is going to be next, and the Z is going to be after that. The Z axis is going to represent the tools depth but we will come back to that later. Right now let's just focus on the X and Y axis (which represents the right to left and up and down movements of the tool) and see where this point is on graph.

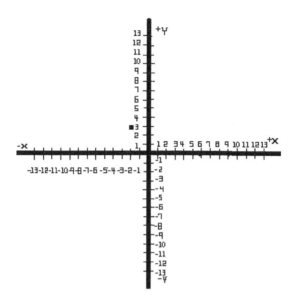

The -2 represents the X axis. It is also negative so you are going to go two lines to the left on the X axis in the negative direction. (If the 2 was positive then you would instead go two lines to the right). Next the Y axis is represented by the 3. The 3 is positive so you are going to go three lines up in the positive direction. (Likewise if the 3 was negative then you would move three lines down in the negative direction). Then the point will be placed at the intersection of the two numbers. Now that we've gotten X and Y out of the way, let's take a look at Z.

Z, like I said before, is going to be represented the tools depth. The origin point for Z will be the top of the part. Anything above the top of the part will be positive, and anything going into the part will be negative. It is represented by a 1 in the set of coordinates above.

One is positive which means it is going to be one of line above the part, and like the others if it was negative it would go down in a negative direction which would be one line into the part.

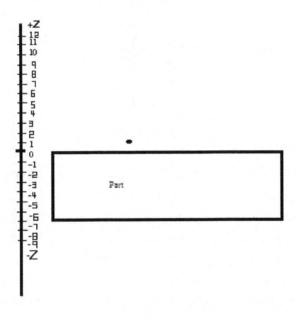

Absolute and Incremental

Absolute and Incremental in the Cartesian Coordinate System

Now that you're familiar with how to use the Cartesian coordinate system the next thing you must determine is which style of programming you wish to use. There is absolute programming where all points originate from the origin point and incremental programming which the origin point changes to the location of the point that you are currently at. I'll show you examples of both styles so you can pick the one that you're most comfortable with.

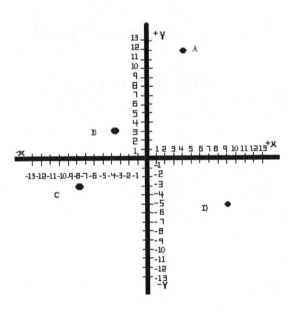

Absolute		
Points	**X**	**Y**
A	4	12
B	-3	3
C	-8	-3
D	9	-5

Incremental		
Points	**X**	**Y**
A	4	12
B	-7	-9
C	-5	-6
D	17	-2

As you can see in absolute programming all points of reference come from a single origin point. In incremental programming the first point of referenced is the origin point then that point becomes your new point of reference. The next point is then calculated from that point instead of the origin. Next let's see how this works in Z.

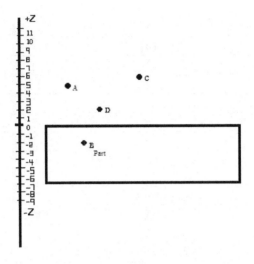

Absolute	
Point	Z
A	5
B	-2
C	6
D	2
Incremental	
Point	Z
A	5
B	-7
C	8
D	-4

As you can see Z is affected in the same way. In absolute mode all points are calculated from the origin point. And in incremental mode the next point is calculated from the last one.

Codes

Letter Address

Next I want to talk about the letters addressed for each code. To specify a certain action a letter address must be placed with a number to tell the machine which action you wish to take place, in some cases how to perform an action. There are also certain symbols that you need to know. So let's take a look at the ones you need to know.

The letter **O** is used to identify the program number. The program number is found at the top of the program on the first line and helps the machine to reference that program.

The letter **N** represents the line number, or in other words what block of programming that you are currently looking at. Even though N addresses are not required they are useful for quickly accessing and keeping track of areas which you wish to change.

The letter **T** is used in the process of changing tools. T along with the tool number will call up the tool in the magazine of the machine.

The letter **M** is for miscellaneous functions such as turning things on and off. Only one M code per block is allowed.

The letter **G** is known as a preparatory code. It will usually put the machine in a specific mode. There can be multiple G codes on a line as long as they do not contradict with each other. For instance you cannot have a G1 and G2 on the same line.

The letter **I** is a code that represents the X axis incrementally from the starting point of a radius to the center of a radius during a radial movement. (G02 or G03).

The letter **J** is a code that represents the Y axis incrementally from the starting point of a radius to the center of the radius during a radial movement. (G02 or G03)

The letter **H** is used to identify which tool height offset to use when G43 is active.

The letter **D** is used to identify which tool diameter offsets to use when cutter compensation is active.

The letter **S** is used to designate the spindle speed.

The letter **F** is used to designate feed rate.

The letter **Q** is used to designate the incremental depth that a drill well travel before backing off to break the chip. It is used with a G83.

The letter **P** is used to designate the time in milliseconds that the tool will pause when using a G04

Parentheses ( ) surrounding text will allow you to write notes to yourself about the program. The machine will not read anything in parentheses. You can use this to identify the start of a tool operation or at the beginning of the program to remind yourself how to set the part in a fixture.

When you place a slash / in front of a line of programming and with the block delete button activated this will cause the machine to skip that line of programming.

The semicolon ; is the last code for each block of programming. It is what concludes for each block of programming. It is known as EOB or end of block.

The percent sign % is the signal for the machine that this is the end of program.

Mobal and Nonmodal

Now that you know how to use the Cartesian coordinate system and you're familiar with letter addresses. I want to talk a little bit about the codes themselves. First you need to understand the difference between modal and nonmodal codes. Modal codes are codes that once given will stay active throughout the program until they are given another code to cancel them such as G1 for linear movements. Nonmodal codes are codes that will only stay active for one line of programming such as G04 for dwell.

G code

G codes are used in correlation with the Cartesian coordinate system to tell the machine what to do. Basically the Cartesian coordinate system tells where the machine to go and G codes tell it how to get there and what to do when it gets there.

G code

Code	Description
G00	Rapid traverse
G01	Linear movement with a feedrate
G02	Clockwise movement with a federate
G03	Counter Clockwise movement with a federate
G04	Dwell
G20	Programming in an inch format
G28	Return to machine the home zero
G40	Tool cutter compensation cancel
G41	Tool cutter compensation left
G42	Tool cutter compensation right
G43	Reference tool height offset
G54-G59	Part home position coordinates preset
G80	Canned cycle cancel
G81	Drilling canned cycle
G83	Pick drilling canned cycle
G90	Absolute mode
G91	Incremental mode
G98	Initial level return
G99	R level return

M codes

Code	Description
M00	Program stop
M01	Optional program stop
M02	End of program
M03	Spindle start clockwise
M05	Spindle stop
M06	Tool change
M08	Flood coolant on
M09	Coolant off
M30	End of program and restart to beginning of program

Now of course there are more codes than this but these are the ones that I'm going to be using in this book. And even though codes are supposed to be standardized, they do tend to vary from machine to machine. However most of the codes I'm going to be using do tend to stay the same. But please check the manual for your machine before using any codes from this book.

Okay now that we've got all of the pieces to the puzzle, let's start putting a few together. In this next section I'm going to actually construct some programs and go line by line explaining what I'm doing and why. First I'm going to do all programs in absolute mode then I'm going to turn around and do all again in incremental that way you can decide which style of programming you prefer.

Absolute

G01 linear movements

This first program is going to use G01 (which is a linear movement) to simply show you how to trace a tool path. What it is going to do is go into the part -.075 and trace the outline below. After that's done it will return to machine home zero. (Just think of it as a game of connect the dots)

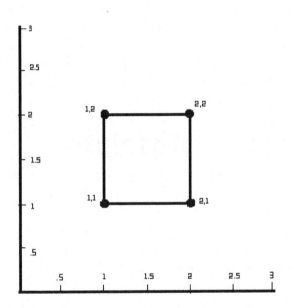

N10 O0001

N20 (Make sure tool one is a half inch spot drill);

N30 (Engraving with 1/2 inch spot drill);

N40 G90 G40 G80 G20;

N50 G0 G28 Z0;

N60 X0 Y0;

N70 T1;

N80 M6;

N90 G54 X1. Y1. S1500 M3;

N100 G43 H1 Z.100 M8 G4 P500;

N110 G1 Z-.075 F80.;

N120 X1. Y2.;

N130 X2. Y2.;

N140 X2. Y1.;

N150 X1. Y1.;

N160 G0 G28 Z0 M5;

N170 X0 Y0 M9;

N180 M30;

%

(N10 O0001)- . N10 is the line number. It goes up by tens so that if you need to add another line of programming you can do so and keep it in order without having to redo all your line numbers. **O0001** This is the program number it helps the machine to identify the program.

(N20 (Make sure tool one is a half inch spot drill) ;)- .
N20 is the line number. **(Make sure tool one is a half inch spot drill)** This is an example of a note that you can leave for yourself to show what you need to do in preparation of running a program. You don't have to worry about it messing the code up because the machine will not read the note as long as it is within parenthesis. The ; is put at the end of every line of programming as a way to show that it is the end of the block of programming.

(N30 (Engraving with 1/2 inch spot drill) ;)- N30 is the line number. **(Engraving with 1/2 inch spot drill)** This is another example of a note you can leave yourself to identify the beginning of an operation. ; End the block.

(N40 G90 G40 G80 G20 ;)- (Note) This first line of programming contains code that is not necessary at this time but is put in as a safety precaution. The only code that is necessary for the machine to function properly is G90. **N40** is the line number. **G90** puts the machine in absolute mode. **G40** is cancels cutter compensation. **G80** is cancels canned cycles. **G20** is use an inch format. Usually the machine will already be sent in the parameters to default to either an inch or metric format but I include this code just in case someone runs this inch formatted program on a machine that is set with a metric default. The code will temporarily put the machine in an inch format until it reads the end of the program then it will go back to its default settings (G21 is the code to put it in a metric format). ; End the block.

(**N50 G0 G28 Z0 ;)- N50** is the line number. **G0** will put the machine in rapid traverse which means the machines axial movements will travel as quickly as they can. It is a modal code which means it will stay in this mode until given another code in the same family. **G28** is the code for it to use the machine home position coordinates. It is a modal code and means that you are going to be working off of the machine home coordinates. **Z0** means it will move Z to the machine home zero position. You will usually move the Z axis back first to avoid a collision. ; End of block.

(**N60 X0 Y0 ;)- N60** is the line number. **X0** means it will move X to the machine home zero. **Y0** means it will move Y to the machine home zero. ; End of block.

(**N70 T1 ;)- N70** is the line number. **T1** calls up the tool in the machine's magazine. (Note) if you wish to call up a different tool number simply put T and the tool number you wish to call up. ; End of block.

(**N80 M6 ;)- N80** is the line number. **M6** activates a tool change. ; End of block.

(**N90 G54 X1. Y1. S1500 M3 ;)- N90** is the line number.**G54** is the code for it to use part home position coordinates in this case it is located on the corner of the part. It is a modal code and means that you are going to be working on the part home coordinates and will also cancel out the G28. **X1.** means it is going to rapid to the X one point inch coordinate. **Y1.** means it is going to rapid to the Y one point inch coordinate. (Note) There both going to move at the same time and they are both going to rapid because G0 is still active from earlier in the program. **S1500** is the speed at which the spindle is going to turn. **M3** is the code for turning on the spindle in a clockwise direction. (Note) the spindle speed and M3 have to be on the same line of programming. **;** End of block.

(N100 G43 H1 Z.100 M8 G4 P500 ;)- **N100** is the line number. **G43** is the code for use the preset tool length offset in the machine. **H1** tells the machine which tool length offset to use, in this case tool one. (Note) G43 is a modal code and will stay active until given a G49 to cancel it. **Z.100** means the tool will rapid and stop point one hundred thousandths above the part. (Note) you want to make sure to start just above the part then go into it with a set feed. **M8** turns on the coolant.(Note) M8 is a modal code and will stay on until canceled with a M9. **G4** is the code for dwell which means the machine will temporarily stop where it's at. This is used when you need to give the machine a little bit of time to catch up with itself. For instance if the coolant pumps are a tad slow and the machine starts cutting before it actually starts spraying coolant, this will allow a little extra time to pass to let the pumps catch up with the machine. **P500** tell the machine how much time in milliseconds to let the machine set before starting up again when using G4. (Note) there are 1,000 milliseconds in a second which means the 500 milliseconds are equal to about ½ seconds. ; End of block.

(**N110 G1 Z-.075 F80. ;)**- **N110** is the line number. **G1** is a linear movement with a feed. This is a modal code and will cancel out the G0. **Z-.075** means it will move into the part negative point zero seventy-five thousandths. **F80.** is the set feed at which the tool will move, which is 80 inches per min. ; End of block.

(**N120 X1. Y2. ;)**- **N120** is the line number. **X1.** it will stay at coordinate one point inch in X. **Y2.** it will move to coordinate two point inches in Y. It will use the same feed rate and linear movement as the previous line. ; End of block.

(**N130 X2. Y2. ;)**- **N130** is the line number.**X2.** it will move to coordinate two point inches in X using the same feed and movement. **Y2.** It will stay at coordinate two point inches in Y. ; End of block.

(**N140 X2. Y1. ;)**- **N140** is the line number. **X2.** it will stay at coordinate two point inches in X. **Y1.** It will move to coordinate one point inch in Y using the same feed and movement. ; End of block.

(**N150 X1. Y1. ;)**- **N150** is the line number. **X1.** it will move to coordinate one point inch in X using the same feed and movement. **Y1.** it will stay at coordinate one point inch in Y. ; End of block.

(N160 G0 G28 Z0 M5 ;)- N160 is the line number. **G0** is rapid traverse. **G28** machine home coordinates. **Z0** is going to move to the 0 coordinate in the machine home position. **M5** will turn off the spindle. ; End of block.

(N170 X0 Y0 M9 ;)- N170 is the line number. **X0** is going to move to machine home zero position in rapid traverse. **Y0** is going to move to machine home zero position in rapid traverse. **M9** is going to turn off the coolant. ; End of block.

(N180 M30 ;)- N180 is the line number. **M30** end of program and then reset itself at beginning of the program. ; End of block.

(%)- % This symbol is given at the end of every program to let the machine know that this is the end of the program. This symbol must be at the end of every program.

G41 cover compensation left and G42 cutter compensation right

In these two programs, I'm going to show you how to use cutter compensation. These are very useful when you want to cut the profile of the part without having to do a large amount of calculations. But first you must decide which of the 2 compensations you should use right or left. To determine these you most put yourself as you were right behind the tool as it is moving forward this will help you determine which of the two you must use. If you will look at the pictures below and place yourself behind the tool you can easily determine which of the two compensations to use. If you look behind the tool as it is moving forward and the tool needs to move to the left to clear the part then you need to use left compensation. Likewise if it's moving forward and it needs to move to the right to clear the part you need to use right compensation.

Right compensation

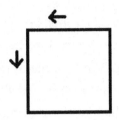

Left compensation

G41 cover compensation left

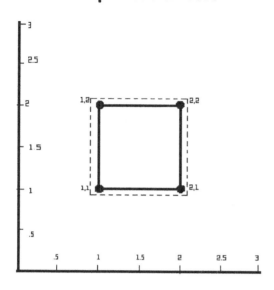

N10 O0002

N20 (Make sure tool two is 1/2 inch end mill);

N30 (Milling profile with half inch end mill);

N40 G90 G40 G80 G20;

N50 G0 G28 Z0;

N60 X0 Y0;

N70 T2;

N80 M6;

N90 G54 G41 D2 X1. Y.5 S1500 M3;

N100 G43 H2 Z.100 M8;

N110 G1 Z-.075 F80.;

N120 X1. Y2.;

N130 X2.Y2.;

N140 X2. Y1.;

N150 X.5 Y1.;

N160 G0 G40 G28 Z0 M5;

N170 X0 Y0 M9;

N180 M2;

%

(N10 O0002)- N10 line number. **O0002** Program number.

(N20 (Make sure tool two is 1/2 inch end mill) ;)- N20 line number. **(Make sure tool two is 1/2 inch end mill)** A note. ; End of block.

(N30 (Milling profile with half inch end mill) ;)- N30 line number. **(Milling profile with half inch end mill)** A note. ; End of block.

(N40 G90 G40 G80 G20 ;)- N40 line number. **G90** puts it in absolute mode. **G40** cancels cutter compensation. **G80** cancels canned cycle. **G20** puts in inch mode. ; End of block.

(N50 G0 G28 Z0 ;)- N50 line number. **G0** is rapid traverse. **G28** use machine home zero coordinates. **Z0** Go to Z zero. ; End the block.

(N60 X0 Y0 ;)- N60 line number. **X0** go to X zero. **Y0** go to Y zero. ; End of block.

(N70 T2 ;)- N70 line number. **T2** call up tool two in the machine magazine. ; End of block.

(N80 M6 ;)- N80 line number. **M6** Preforms tool change. ; End of block.

(N90 G54 G41 D2 X1. Y.5 S1500 M3 ;)- N90 line number. **G54** use part home zero coordinates. **G41** activates cutter compensation left. **D2** is used tool diameter offsets for tool two. **X1.** go to X one. **Y.5** goes to Y point five. (Note) when preforming a milling operation it is better to start off the part then feed into it. **S1500** is spindle speed. **M3** turn on spindle. ; End of block.

(N100 G43 H2 Z.100 M8 ;)- N100 line number.**G43** use tool offsets. **H2** used tool height offset for tool two. **Z.100** it will move to point hundred above the part. **M8** turn on coolant. ; End of block.

(N110 G1 Z-.075 F80. ;)- N110 line number. **G1** is a linear movement. **Z-.075** it will move to negative point zero seventy-five into the part. **F80.** feed rate at eighty. ; End of block.

(N120 X1. Y2. ;)- N120 line number. **X1.** stay at X one.**Y2.** move to Y two. ; End of block.

(N130 X2. Y2. ;)- N130 line number. **X2.** move to X two. **Y2.** stay at Y two. ; End of block.

(N140 X2. Y1. ;)- N140 line number. **X2.** stay at X two point inches. **Y1.** move to Y one. ; End of block.

(N150 X.5 Y1. ;)- N150 line number. **X.5** move to X point five. (Note) when preforming a milling operation it is better to move the tool all the way off the part when finished. **Y1.** stay at Y one. ; End of block.

(N160 G0 G40 G28 Z0 M5 ;)- N160 line number. **G0** is rapid traverse. **G40** is cancels cutter compensation. **G28** use machine home zero position coordinates. **Z0** move to Z zero. **M5** turn off spindle. ; End of block.

(N170 X0 Y0 M9 ;)- N170 line number. **X0** move to X zero. **Y0** move to Y zero. **M9** turn coolant off. ; End of block.

(N180 M2 ;)- N180 line number. **M2** end of program. ; End of block.

(%)- % End of the program.

G42 cutter compensation right

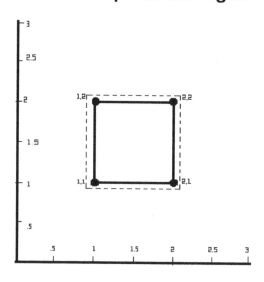

N10 O0003

N20 (Make sure tool two is 1/2 inch end mill);

N30 (Milling profile with half inch end mill);

N40 G90 G40 G80 G20;

N50 G0 G28 Z0;

N60 X0 Y0;

N70 T2;

N80 M6;

N90 G54 G42 D2 X.5 Y1 S1500 M3;

N100 G43 H2 Z.100 M8;

N110 G1 Z-.075 F80.;

N120 X2. Y1.;

N130 X2. Y2.;

N140 X1. Y2.;

N150 X1. Y.5;

N160 G0 G40 G28 Z0 M5;

N170 X0 Y0 M9;

M180 M2;

%

(N10 O0003)- N10 line number. **O0003** Program number.

(N20 (Make sure tool two is 1/2 inch end mill) ;)- N20 line number. **(Make sure tool two is 1/2 inch end mill)** A note. ; End of block.

(N30 (Milling profile with half inch end mill) ;)- N30 line number. **(Milling profile with half inch end mill)** A note. ; End of block.

(N40 G90 G40 G80 G20 ;)- N40 line number. **G90** puts it in absolute mode. **G40** is cancels cutter compensation. **G80** is cancels canned cycle. **G20** puts it in inch mode. ; End of block.

(N50 G0 G28 Z0 ;)- N50 line number.**G0** is rapid traverse. **G28** use machine home position zero coordinates. **Z0** go to Z zero. **;** End the block.

(N60 X0 Y0 ;)- N60 line number. **X0** go to X zero.**Y0** go to Y zero. **;** End of block.

(N70 T2 ;)- N70 line number. **T2** call up tool two in the machine magazine. **;** End of block.

(N80 M6 ;)- N80 line number. **M6** performs a tool change. **;** End of block.

(N90 G54 G42 D2 X.5 Y1. S1500 M3 ;)- N90 line number. **G54** use part home zero position coordinates. **G42** activates cutter compensation right. **D2** used tool diameter offsets for tool two. **X.5** go to X point five. (Note) when preforming a milling operation and is better to start off the part then feed into it. **Y1.** go to Y one point inch. **S1500** spindle speed. **M3** turn on spindle. **;** End of block.

(N100 G43 H2 Z.100 M8 ;)- N100 line number.**G43** tool offsets. **H2** used tool height offset for tool two. **Z.100** moves to point hundred above the part. **M8** turn on coolant. **;** End of block.

(N110 G1 Z-.075 F80. ;)- N110 line number. **G1** is a linear movement. **Z-.075** move to negative point zero seventy-five into the part. **F80.** set feed rate at eighty. **;** End of block.

(N120 X2. Y1. ;)- **N120** line number. **X2.** move to X two.**Y1.** stay at Y one. ; End of block.

(N130 X2. Y2. ;)- **N130** line number. **X2.** stay at X two. **Y2.** move to Y two. ; End of block.

(N140 X1. Y2. ;)- **N140** line number. **X1.** move to X one. **Y2** stay in Y two. ; End of block.

(N150 X1. Y.5 ;)- **N150** line number. **X1.** stay at X one. **Y.5** move to Y point five. (Note) when preforming a milling operation is better to move the tool all the way off the part when finished.. ; End of block.

(N160 G0 G40 G28 Z0 M5 ;)- **N160** line number. **G0** is rapid traverse. **G40** is cancels cutter compensation. **G28** use machine home zero position coordinates. **Z0** move to Z zero. **M5** turn off spindle. ; End of block.

(N170 X0 Y0 M9 ;)- **N170** line number. **X0** moved to X zero. **Y0** move to Y zero. ; End of block. **M9** turn coolant off.

(M180 M2 ;)- **N180** line number. **M2** end of program. ; End of block.

(%)- **%** End of the program.

G02 and G03 with R

G02 and G03 are radial movements which means they will move from one point to another and form of radius as they go. There are two ways of doing this, using R along with the radius measurement of the arc you wish to create is the simpler of the two ways. However it cannot be used for all circles or arcs only those that are 180° or under. If you don't know what that means then think of it this way a circle has 360° half of that would be 180°. So this method will work on any that is a half of a circle or less.

G02 with R

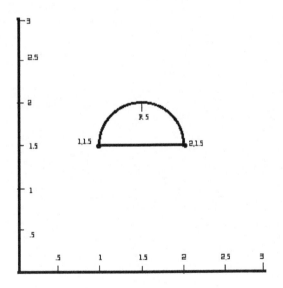

N10 O0004

N20 (Make sure tool one is a half inch spot drill);

N30 (Engraving with 1/2 inch spot drill);

N40 G90 G40 G80 G20;

N50 G0 G28 Z0;

N60 X0 Y0;

N70 T1;

N80 M6;

N90 G54 X1. Y1.5 S1500 M3;

N100 G43 H1 Z.100 M8;

N110 G1 Z-.075 F80.;

N120 G2 X2. Y1.5 R.5;

N130 G1 X1. Y1.5;

N140 G0 G28 Z0 M9;

N150 X0 Y0 M5;

N160 M30;

%

(N10 O0004)- **N10** line number. **O0004** Program number.

(N20 (Make sure tool two is 1/2 inch spot drill) ;)- **N20** line number. **(Make sure tool two is 1/2 inch spot drill)** A note. ; End of block.

(N30 (Milling profile with half inch spot drill) ;)- **N30** line number. **(Milling profile with half inch spot drill)** A note. ; End of block.

 (N40 G90 G40 G80 G20 ;)- **N40** line number. **G90** puts it in absolute mode. **G40** is cancels cutter compensation. **G80** is cancels canned cycle. **G20** puts it in inch mode. ; End of block.

(N50 G0 G28 Z0 ;)- **N50** line number. **G0** is rapid traverse. **G28** use machine home position coordinates. **Z0** move to Z zero. ; End of block.

(N60 X0 Y0 ;)- **N60** line number.**X0** move to X zero. **Y0** move to Y zero. ; End of block.

(N70 T1 ;)- **N70** line number. **T1** call up tool one in magazine. ; End of block.

(N80 M6 ;)- **N80** line number. **M6** preform tool change. ; End of block.

(N90 G54 X1. Y1.5 S1500 M3 ;)- **N90** line number. **G54** use part home position coordinates. **X1.** move to X one.**Y1.5** move to Y one point five. **S1500** spindle speed. **M3** turn on spindle. ; End of block.

(N100 G43 H1 Z.100 M8 ;)- N100 line number. **G43** used tool length offset. **H1** use height offset for tool one. **Z.100** move to point one hundred above the part. **M8** turn on coolant. ; End of block.

(N110 G1 Z-.075 F80. ;)- N110 line number. **G1** is linear movement. **Z-.075** move to negative point zero seventy-five into the part. **F80.** set feed rate. ; End of block.

(N120 G2 X2. Y1.5 R.5 ;)- N120 line number. **G2** this is a movement that will give a clockwise radius. (Note) G2 is a modal code and will remain active unless given another code to cancel it. If the feed rate is specified earlier in the program with a G01 the feed rate will remain the same otherwise specify a feedrate when you use a G02 or G03. **X2.** move to X two. **Y1.5** stay at Y one point five. **R.5** the arc has a radius of point five. ; End of block.

(N130 G1 X1. Y1.5 ;)- N130 line number. **G1** is a linear movement. **X1.** move to X one. **Y1.5** stay at Y one point five. ; End of block.

(N140 G0 G28 Z0 M9 ;)- N140 line number. **G0** is rapid traverse. **G28** use machine home position coordinates. **Z0** move to Z zero. **M9** turn off the coolant. ; End of block.

(N150 X0 Y0 M5 ;)- N150 line number. **X0** move to X zero. **Y0** move to Y zero. **M5** turn off spindle. ; End of block.

(N160 M30 ;)- N160 line number. **M30** end of program then reset program. ; End of block.

(%)- % End of the program.

G03 with R

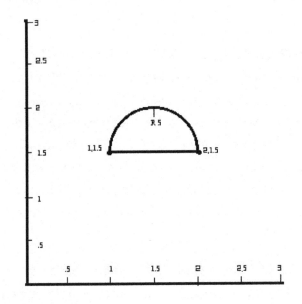

N10 O0005

N20 (Make sure tool one is a half inch spot drill);

N30 (Engraving with 1/2 inch spot drill);

N40 G90 G40 G80 G20;

N50 G0 G28 Z0;

N60 X0 Y0;

N70 T1;

N80 M6;

N90 G0 G54 X1. Y1.5 S1500 M3;

N100 G43 H1 Z.100 M8;

N110 G1 Z-.075 F80.;

N120 G1 X2. Y1.5;

N130 G3 X1. Y1.5 R.5;

N140 G0 G28 Z0 M9;

N150 X0 Y0 M5;

N160 M30;

%

(N10 O0005)- N10 line number. **O0005** Program number.

(N20 (Make sure tool two is 1/2 inch spot drill) ;)- N20 line number. **(Make sure tool two is 1/2 inch spot drill)** A note. **;** End of block.

(N30 (Milling profile with half inch spot drill) ;)- N30 line number. **(Milling profile with half inch spot drill)** A note. **;** End of block.

(N40 G90 G40 G80 G20;)- N40 line number. **G90** puts it in absolute mode. **G40** is cancels cutter compensation.

G80 is cancels canned cycle. **G20** puts it in inch mode. ; End of block.

(N50 G0 G28 Z0 ;)- **N50** line number. **G0** is rapid traverse. **G28** use machine home position coordinates. **Z0** move to Z zero. ; End of block.

(N60 X0 Y0 ;)- **N60** line number. **X0** move to X zero. **Y0** move to Y zero. ; End of block.

(N70 T1 ;)- **N70** line number. **T1** call up tool one in magazine. ; End of block.

(N80 M6 ;)- **N80** line number. **M6** preform tool change. ; End of block.

(N90 G54 X1. Y1.5 S1500 M3 ;)- **N90** line number. **G54** use part home position coordinates. **X1.** move to X one. **Y1.5** move to Y one point five. **S1500** spindle speed. **M3** turn on spindle. ; End of block.

(N100 G43 H1 Z.100 M8 ;)- **N100** line number. **G43** used tool length offset. **H1** use height offset for tool one. **Z.100** move to point one hundred above the part. **M8** turn on coolant. ; End of block.

(N110 G1 Z-.075 F80. ;)- **N110** line number. **G1** is linear movement. **Z-.075** move to negative point zero seventy-five into the part. **F80.** set feed rate. ; End of block.

(N120 G1 X2. Y1.5 ;)- **N120** line number. **G1** is a linear movement. **X2.** move to X two. **Y1.5** stay at Y one point five. ; End of block.

(N130 G3 X1. Y1.5 R.5 ;)- **N130** line number. **G3** this is a movement that will give a counter clockwise radius. (Note) G3 is a modal code and will remain active unless given another code to cancel it. If the feed rate is specified earlier in the program with a G01 the feed rate will remain the same unless otherwise specified when you use a G02 or G03. **X1.** move to X one. **Y1.5** stay at Y one point five. **R.5** the arc has a radius of point five. ; End of block.

(N140 G0 G28 Z0 M9 ;)- **N140** line number. **G0** is rapid traverse.**G28** use machine home position coordinates. **Z0** move to Z zero. **M9** turn off the coolant. ; End of block.

(N150 X0 Y0 M5 ;)- **N150** line number. **X0** move to X zero. **Y0** move to Y zero. **M5** turn off spindle. ; End of block.

(N160 M2 ;)- **N160** line number. **M30** end of program then reset program. ; End of block.

(%)- **%** End of the program.

G02 and G03 with I and J

Using I and J is the second method of using G02 or G03. It can be used on any degree of arc or circle but does take a little more work to use. To use I and J first you must move the tool to the starting point of the arc or circle you wish to create. Then you must determine in an incremental style (even when using absolute) the distance between the starting point to the center of the circle. Then using I to represent the X and J to represent Y write the coordinates down along with the ending point to form the arc or circle.

G02 with I and J

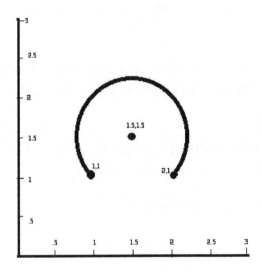

N10 O0006

N20 (Make sure tool one is a half inch spot drill);

N30 (Engraving with 1/2 inch spot drill);

N40 G90 G40 G80 G20;

N50 G0 G28 Z0;

N60 X0 Y0;

N70 T1;

N80 M6;

N90 G0 G54 X1. Y1. S1500 M3;

N100 G43 H1 Z.100 M8;

N110 G1 Z-.075 F80.;

N120 G02 X2. Y1. I.5 J.5;

N130 G0 G28 Z0 M9;

N140 X0 Y0 M5;

N150 M2;

%

(N10 O0006)- N10 line number. **O0006** Program
number.

(N20 (Make sure tool two is 1/2 inch spot drill) ;)- **N20** line number. **(Make sure tool two is 1/2 inch spot drill)** A note. **;** End of block.

(N30 (Milling profile with half inch spot drill) ;)- **N30** line number. **(Milling profile with half inch spot drill)** A note. **;** End of block.

 (N40 G90 G40 G80 G20 ;)- **N40** line number. **G90** puts it in absolute mode. **G40** is cancels cutter compensation. **G80** is cancels canned cycle. **G20** puts in inch mode. **;** End of block.

(N50 G0 G28 Z0 ;)- **N50** line number. **G0** is rapid traverse. **G28** use machine home position coordinates. **Z0** move to **Z** zero. **;** End of block.

(N60 X0 Y0 ;)- **N60** line number.**X0** move to X zero. **Y0** move to Y zero. **;** End of block.

(N70 T1 ;)- **N70** line number. **T1** call up tool one in magazine. **;** End of block.

(N80 M6 ;)- **N80** line number. **M6** preform tool change. **;** End of block.

(N90 G54 X1. Y1. S1500 M3 ;)- **N90** line number. **G54** use part home position coordinates. **X1.** move to X one.**Y1.** move to Y one. **S1500** spindle speed. **M3** turn on spindle. **;** End of block.

(N100 G43 H1 Z.100 M8 ;)- **N100** line number. **G43** used tool length offset. **H1** use height offset for tool one. **Z.100** move to point one hundred above the part. **M8** turn on coolant. ; End of block.

(N110 G1 Z-.075 F80. ;)- N110 line number. **G1** is linear movement. **Z-.075** move to negative point zero seventy-five into the part. **F80.** set feed rate. ; End of block.

(N120 G2 X2. Y1. I.5 J.5 ;)- N120 line number. **G2** it will move in a clockwise direction. **X2.** go to X two. **Y1.** stay at Y one. **I.5** represents the distance between the starting point of the cut to the center of the arc or circle in X. The distance is always calculated in an incremental format even in absolute mode. **J.5** represents the distance between the starting point of the cut to the center of the arc or circle in Y. The distance is always calculated in an incremental format even in absolute mode. ; End of block.

(N130 G0 G28 Z0 M9 ;)- N130 line number. **G0** is rapid traverse.**G28** use machine home position coordinates. **Z0** move to Z zero. **M9** turn off the coolant. ; End of block.

(N140 X0 Y0 M5 ;)- N140 line number. **X0** move to X zero. **Y0** moved to Y zero. **M5** turn off spindle. ; End of block.

(N150 M2 ;)- N150 line number. **M2** end of program. ; End of block.

(%)- % End of the program.

G03 with I and J

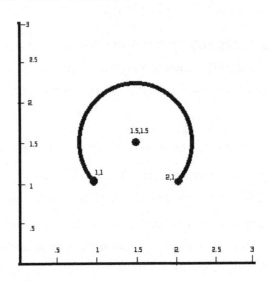

N10 O0007

N20 (Make sure tool one is a half inch spot drill);

N30 (Engraving with 1/2 inch spot drill);

N40 G90 G40 G80 G20;

N50 G0 G28 Z0;

N60 X0 Y0;

N70 T1;

N80 M6;

N90 G0 G54 X2. Y1. S1500 M3;

N100 G43 H1 Z.100 M8;

N110 G1 Z-.075 F80.;

N120 G03 X1. Y1. I -.5 J.5;

N130 G0 G28 Z0 M9;

N140 X0 Y0 M5;

N150 M2;

%

(N10 O0007)- N10 line number. **O0007** Program number.

(N20 (Make sure tool two is 1/2 inch spot drill) ;)- N20 line number. **(Make sure tool two is 1/2 inch spot drill)** A note. ; End of block.

(N30 (Milling profile with half inch spot drill) ;)- N30 line number. **(Milling profile with half inch spot drill)** A note. ; End of block.

(N40 G90 G40 G80 G20 ;)- N40 line number. **G90** puts it in absolute mode. **G40** is cancels cutter compensation. **G80** is cancels canned cycle. **G20** puts it in inch mode. ; End of block.

(N50 G0 G28 Z0 ;)- N50 line number. **G0** is rapid traverse. **G28** use machine home position coordinates. **Z0** move to Z zero. ; End of block.

(N60 X0 Y0 ;)- N60 line number.**X0** move to X zero. **Y0** move to Y zero. ; End of block.

(N70 T1 ;)- **N70** line number. **T1** call up tool one in magazine. ; End of block.

(N80 M6 ;)- **N70** line number. **M6** preform tool change. ; End of block.

(N90 G54 X2. Y1. S1500 M3 ;)- **N90** line number. **G54** use part home position coordinates. **X2.** move to X two.**Y1.** move to Y one. **S1500** spindle speed. **M3** turn on spindle. ; End of block.

(N100 G43 H1 Z.100 M8 ;)- **N100** line number. **G43** used tool length offset. **H1** use height offset for tool one. **Z.100** move to point one hundred above the part. **M8** turn on coolant. ; End of block.

(N110 G1 Z-.075 F80. ;)- **N110** line number. **G1** is linear movement. **Z-.075** move to negative point zero seventy-five into the part. **F80.** set feed rate. ; End of block.

(N120 G3 X1. Y1. I -.5 J.5 ;)- **N120** line number. **G3** it will move in a counter clockwise direction. **X1.** go to X one. **Y1.** stay at Y one. **I -.5** I represents the distance between the starting point of the cut to the center of the arc or circle in X. Distance is always calculated in an incremental format even in absolute mode. **J.5** J represents the distance between the starting point of the cut to the center of the arc or circle in Y. The distance is always calculated in an incremental format even in absolute mode. ; End of block.

(N130 G0 G28 Z0 M9 ;)- **N130** line number. **G0** is rapid traverse.**G28** use machine home position coordinates. **Z0** move to **Z** zero. **M9** turn off the coolant. ; End of block.

(N140 X0 Y0 M5 ;)- **N140** line number. **X0** move to X zero. **Y0** move to Y zero. **M5** turn off spindle. ; End of block.

(N150 M2 ;)- **N150** line number. **M2** end of program. ; End of block.

(%)- % End of the program.

Using canned drill cycles

      Using a canned drilling cycle makes drilling a large number of holes quite easy since you do not have to reenter the information about the hole at each point. You simply enter the information once then simply give it the next location you wish to the drill.

Canned drilling cycles

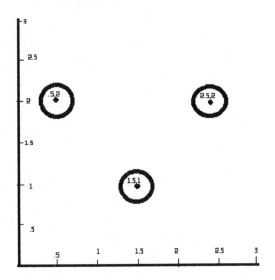

N10 O0008

N20 (Tool 1 needs to be 1/2 inch spot drill);

N30 (Tool 3 needs to be a 27/64 inch drill);

N40 (Tool1 Spot Drill 1/2 inch);

N50 G90 G40 G80 G20;

N60 G0 G28 Z0;

N70 X0 Y0;

N80 T1;

N90 M6;

N100 G54 X.5 Y2. S1500 M3;

N110 G43 H1 Z.100 M8 G4 P500;

N120 G81 G99 Z-.025 R.100 F80.;

N130 X1.5 Y1.;

N140 X2.5 Y2.;

N150 G80;

N160 G0 G28 Z0 M5;

N170 X0 Y0 M9;

N180 M1;

N190 (Tool3 Drill 27/64);

N200 G90 G40 G80 G20;

N210 G0 G28 Z0;

N220 X0 Y0;

N230 T3;

N240 M6;

N250 G54 X.5 Y2. S1500 M3;

N260 G43 H3 Z.100 M8 G4 P500;

N270 G83 G98 Z-.500 Q.100 R.100 F80.;

N280 / X1.5 Y1.;

N290 / X2.5 Y2.;

N300 G80;

N310 G0 G28 Z0 M5;

N320 X0 Y0 M9;

N330 M1;

N340 (Tool1 Chamfer 1/2 Spot Drill);

N350 G90 G40 G80 G20;

N360 G0 G28 Z0;

N370 X0 Y0;

N380 T1;

N390 M6;

N400 G54 X.5 Y2. S1500 M3;

N410 G43 H1 Z.100 M8 G4 P500;

N420 G81 G99 Z-.075 R.100 F80.;

N430 X1.5 Y1.;

N440 X2.5 Y2.;

N450 G80;

N460 G0 G28 Z0 M5;

N470 X0 Y0 M9;

N480 M2;

%

(N10 O0008 ;)- **N10** line number. **O0008** Program number.

(N20 (Tool 1 needs to be 1/2 inch spot drill) ;)- **N20** line number. **(Tool 1 needs to be 1/2 inch spot drill)** A note. ; End the block.

(N30 (Tool 3 needs to be a 27/64 inch drill) ;) – **N30** line number. **(Tool 3 needs to be a 27/64 inch drill)** A note. ; End of block.

(N40 (Tool 1 Spot Drill 1/2 inch) ;)- **N40** line number. **(Tool1 Spot Drill 1/2 inch)** A note. ; End of block.

(N50 G90 G40 G80 G20 ;)- **N50** line number. **G90** puts it in absolute mode. **G40** is cancels cutter compensation. **G80** is cancel canned cycle. **G20** puts it in inch format. ; End of block.

(N60 G0 G28 Z0 ;)- **N60** line number. **G0** is rapid traverse. **G28** use machine home position coordinates. **Z0** move to Z zero. ; End of block.

(N70 X0 Y0 ;)- N70 line number.**X0** move to X zero. **Y0** move to Y zero. ; End of block.

(N80 T1 ;)- N80 line number. **T1** call up tool one in magazine. ; End of block.

(N90 M6 ;)- N90 line number. **M6** preform tool change. ; End of block.

(N100 G54 X.5 Y2. S1500 M3 ;)- N100 line number. **G54** use part home position coordinates. **X.5** move to **X** point five. **Y2.** move to **Y** two. **S1500** spindle speed. **M3** turn on spindle. ; End of block.

(N110 G43 H1 Z.200 M8 G4 P500 ;)- N110 line number. **G43** used tool length offset. **H1** use height offset for tool one. **Z.200** move to point two hundred above the part. **M8** turn on coolant.**G4** dwell. **P500** the amount of time for dwell. ; End of block.

(N120 G81 G99 Z-.025 R.100 F80. ;)- N120 line number.**G81** is a canned drill cycle. This is a modal code, and will perform the same drilling action that is specified on this line at each location without need of further input on the other lines.**G99** Is R level return. This tells the machine to use R for the return height. **Z-.025** will move negative point zero twenty-five into the part. **R.100** the R in this case stands for retract height (if used with a G99). Which means the height at which it will return after it is done drilling the hole. In this case is .100 above the part. **F80.** set feed rate at eighty. ; End the block.

(N130 X1.5 Y1. ;)- **N130** line number. **X1.5** move to X one point five. **Y1.** move to Y one. ; End of block.

(N140 X2.5 Y2. ;)- **N140** line number. **X2.5** move to X two point five. **Y2.** move to Y two. ; End of block.

(N150 G80 ;)- **N150** line number. **G80** cancels canned cycle. ; End of block.

(N160 G0 G28 Z0 M5 ;)- **N160** line number. **G0** is rapid traverse.**G28** use machine home position coordinates. **Z0** move to **Z** zero. **M5** turn off spindle. ; End of block.

(N170 X0 Y0 M9 ;)- **N170** line number. **X0** move to X zero. **Y0** move to Y zero. **M9** turn off the coolant. ; End of block.

(N180 M1 ;)- **N180** line number. **M1** is optional stop. This will allow you to stop the machine at this point as long as the optional stop button is pressed on the machine. It is good to place an M1 at the end of operations that way you can stop the machine and check the part if needed. It also gives you a way to stop the machine without having to wait for the completion of the program. ; End the block.

(N190 (Tool3 Drill 27/64) ;)- **N190** line number. **(Tool 3 Drill 27/64)** A note. ; End the block.

(N200 G90 G40 G80 G20 ;)- **N200** line number. **G90** puts it in absolute mode. **G40** is cancels cutter compensation. **G80** is cancels canned cycle. **G20** puts it in inch format. **;** End of block.

(N210 G0 G28 Z0 ;)- N210 line number. **G0** is rapid traverse. **G28** use machine home position coordinates. **Z0** move to **Z** zero. **;** End of block.

(N220 X0 Y0 ;)- N220 line number.**X0** move to X zero. **Y0** move to Y zero. **;** End of block.

(N230 T3 ;)- N230 line number. **T3** call up tool three in magazine. **;** End of block.

(N240 M6 ;)- N240 line number. **M6** preform tool change. **;** End of block.

(N250 G54 X.5 Y2. S1500 M3 ;)- N250 line number. **G54** use part home position coordinates. **X.5** move to X point five. **Y2.** move to Y two. **S1500** spindle speed. **M3** turn on spindle. **;** End of block.

(N260 G43 H1 Z.100 M8 G4 P500 ;)- N260 line number. **G43** used tool length offset. **H1** use height offset for tool one. **Z.100** move to point one hundred above the part. **M8** turn on coolant.**G4** dwell. **P500** the amount of time for dwell. **;** End of block.

(N270 G83 G98 Z-.500 Q.100 R.100 F80. ;)- N270 line number.**G83** is a canned drill cycle. This is a modal code, and will perform the same drilling action that is specified on this line at each location without need of further input on the other lines. This code is known as a peck drill cycle. It will feed so far into the part before backing off to break the chip then continuing to drill till it reaches the next target increment or finishes the hole. **G98** is initial level return. It tell the machine to use the last Z height instead of the R return height which is .100. **Z-.500** will move negative point five hundred into the part. **Q.100** Q is how you set the target increment for the Peck drilling cycle in this case it is set to .100 which mean it will perform a chip break every hundred thousandths into the part. **R.100** retracts height point one hundred (if used a G99) but is not used in the operation. F80. set feed rate at eighty. ; End the block.

(N280 / X1.5 Y1. ;)- **N280** line number. **/** When the block delete button is activated this will tell the machine not to read this line of programming. This is useful when you wish to only run one of a particular action so you can check accuracy or quality of the process. When the block delete button is not on, the machine will read this line of programming normally. **X1.5** move to X one point five. **Y1.** move to Y one. ; End of block.

(N290 / X2.5 Y2. ;)- **N290** line number. **/** Skip this line of programming when block delete is active. **X2.5** move to X two point five. **Y2.** move to Y two. ; End of block.

(N300 G80 ;)- **N300** line number. **G80** cancels canned cycle. ; End of block.

(**N310 G0 G28 Z0 M5 ;**)- **N310** line number. **G0** is rapid traverse. **G28** use machine home position coordinates. **Z0** move to Z zero. **M5** turn off spindle. **;** End of block.

(**N320 X0 Y0 M9 ;**)- **N320** line number. **X0** move to X zero. **Y0** move to Y zero. **M9** turn off the coolant. **;** End of block.

(**N330 M1 ;**)- **N330** line number. **M1** is optional stop. **;** End of block.

(**N340 (Tool 1 Chamfer 1/2 Spot Drill) ;**)- **N340** line number. (**Tool 1 Chamfer 1/2 Spot Drill**) A note. **;** End of block.

(**N350 G90 G40 G80 G20 ;**)- **N350** line number. **G90** puts it in absolute mode. **G40** is cancels cutter compensation. **G80** is cancels canned cycle. **G20** use inch format. **;** End of block.

(**N360 G0 G28 Z0 ;**)- **N360** line number. **G0** is rapid traverse. **G28** use machine home position coordinates. **Z0** move to Z zero. **;** End of block.

(**N370 X0 Y0 ;**)- **N370** line number. **X0** move to X zero. Y0 move to Y zero **;** End of block.

(**N380 T1 ;**)- **N380** line number. **T1** call up tool one in magazine. **;** End of block.

(**N390 M6 ;**)- **N390** line number. **M6** preform tool change. **;** End of block.

(N400 G54 X.5 Y2. S1500 M3 ;)- N370 line number. **G54** use part home position coordinates. **X.5** move to X point five. **Y2.** move to Y two. **S1500** spindle speed. **M3** turn on spindle. ; End of block.

(N410 G43 H1 Z.100 M8 G4 P500 ;)- N410 line number. **G43** used tool length offset. **H1** use height offset for tool one. **Z.100** move to point one hundred above the part. **M8** turn on coolant.**G4** dwell. **P500** the amount of time for dwell. ; End of block.

(N420 G81 G99 Z-.075 R.100 F80. ;)- N420 line number.**G81** is a canned drill cycle. **G99** Is R level return. **Z-.075** will move negative point zero seventy-five into the part. **R.100** retract height to point one hundred (if using a G99). **F80.** set feed rate at eighty. ; End of block.

(N430 X1.5 Y1. ;)- N430 line number. **X1.5** move to X one point five. **Y1.** move to Y one. ; End of block.

(N440 X2.5 Y2. ;)- N440 line number. **X2.5** move to X two point five. **Y2.** move to Y two. ; End of block.

(N450 G80 ;)- N450 line number. **G80** cancels canned cycle. ; End of block.

(N460 G0 G28 Z0 M5 ;)- N460 line number. **G0** is rapid traverse.**G28** use machine home position coordinates. **Z0** move to Z zero. **M5** turn off spindle. ; End of block.

(N470 X0 Y0 M9 ;)- **N470** line number. **X0** move to X zero. **Y0** move to Y zero. **M9** turn off the coolant. ; End of block.

(N480 M2 ;)- **N480** line number. **M2** end of program. ; End the block.

(%)- **%** Symbol for in the program

Incremental

G01 linear movements

This first program is going to uses G01 (which is a linear movement) to simple show you how to trace a tool path. What it is going to do is go into the part -.075 and trace the outline below. After its done it will return to machine home zero.

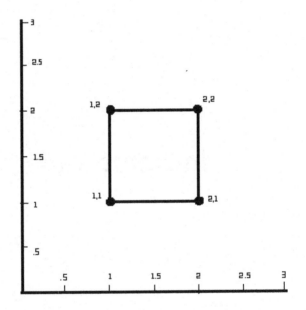

N10 O0001

N20 (Make sure tool one is a half inch spot drill);

N30 (Engraving with 1/2 inch spot drill);

N40 G91 G40 G80 G20;

N50 G0 G28 Z0;

N60 X0 Y0;

N70 T1;

N80 M6;

N90 G54 X1. Y1. S1500 M3;

N100 G43 H1 Z.100 M8 G4 P500;

N110 G1 Z-.175 F80.;

N120 X0 Y1.;

N130 X1. Y0;

N140 X0 Y-1.;

N150 X-1. Y0;

N170 G0 G28 Z0 M5;

N180 X0 Y0 M9;

N190 M2;

%

(N10 O0001)- N10 is the line number. It goes up by tens so that if you need to add another line of programming you can do so and keep them in order without having to redo all your line numbers. **O0001**This line is the program number it helps the machine to identify the program.

(N20 (Make sure tool one is a half inch spot drill) ;)-
N20 is the line number. **(Make sure tool one is a half inch spot drill)** This is an example of a note that you can leave for yourself to show what you need to do in preparation of running a program. You do not have to worry about it messing the code up because the machine will not read the note as long as it is within parenthesis. ; End the block.

(N30 (Engraving with 1/2 inch spot drill) ;)- N30 is the line number. **(Engraving with 1/2 inch spot drill)** This is another example of a note you can leave yourself to identify the beginning of an operation. ; End the block.

(N40 G91 G40 G80 G20 ;)- (Note)This first line of programming contains code that is not necessary at this time but is put in as a safety precaution. The only code that is necessary for the program to function properly is G91. **N40** is the line number. **G91** puts the machine in incremental mode. **G40** is cancels cutter compensation. **G80** is cancels canned cycles. **G20** is use an inch format. Usually the machine will already be sent in the parameters to default to either an inch or metric format but I include this code just in case someone runs this inch formatted program on a machine that is set with a metric default. The code will temporarily put the machine in an inch format until it reads the end of the program then it will go back to its default settings (G21 is the code to put it in a metric format). The ; is put at the end of every line of programming to show that it is the end of block of programing.

(N50 G0 G28 Z0 ;)- N50 is the line number. **G0** will put the machine in rapid traverse which means the machines axial movements will travel as quickly as they came. It is a modal code which means it will stay in this mode until given another code in the same family. **G28** is the code for it to use the machine home position coordinates. It is a modal code and means that you are going to be working off of the machine home coordinates. **Z0** means it will move Z to the machine home zero position. You will usually move the Z axis back first to avoid a collision. ; End of block.

(N60 X0 Y0 ;)- N60 is the line number. **X0** means it will move X to the machine home zero. **Y0** means it will move Y to the machine home zero. ; End of block.

(N70 T1 ;)- N70 is the line number. **T1** calls up the tool one in the machine's magazine. (Note) if you wish to call up a different tool number simply put T and the tool number you wish to call up. ; End of block.

(N80 M6 ;)- N80 is the line number. **M6** activates a tool change. ; End of block.

(N90 G54 X1. Y1. S1500 M3 ;)- N90 is the line number. **G54** is the code for it to use part home position coordinates. It is a modal code and means that you are going to be working on the part home coordinates and will also cancel out the G28. **X1.** means it is going to rapid to X one point inch coordinate.**Y1.** means it is going to rapid to the Y one point inch coordinate. (Note) There both going to move at the same time and they are both going to rapid because G0 is still active from earlier in the program. **S1500** is the speed at which the spindle is going to turn.**M3** is the code for turning on the spindle in a clockwise direction. (Note) the spindle speed and M3 have to be on the same line of programming. ; End of block.

(N100 G43 H1 Z.100 M8 G4 P500 ;)- N100 is the line number. **G43** is the code for use the preset tool length offset in the machine. **H1** tells the machine which tool length offset to use in this case tool one. (Note)G43 is a modal code and will stay active until given a G49 to cancel it. **Z.100** means the tool will rapid and stop at point one hundred above the part. (Note) you want to make sure to start just above the part then go into it at a set feed. **M8** turns on the coolant. (Note) M8 is a modal code and will stay home until canceled with a M9. **G4** is the code for dwell which means the machine will temporarily stop where it's at. This is used for when you need to give the machine a little bit of time to catch up with itself. For instance if the coolant pumps are a tad slow and the machine starts cutting before they actually start spraying coolant. This will allow a little extra time to pass to let the pumps catch up with the machine. **P500** tell the machine how much time in milliseconds to let the machine set before starting up again when using G4. (Note) there are 1,000 milliseconds in a second which means the 500 millisecond is equal to about ½ seconds. ; End of block.

(N110 G1 Z-.175 F80. ;)- N100 is the line number. **G1** is a linear movement with a feed. This is a modal code and will cancel out the G0. **Z-.175** means it will move into the part negative point zero seventy-five. **F80.** is the set feed at which the tool will move which is 80 inches per min. ; End of block.

(N120 X0 Y1. ;)- N120 is the line number. **X0** it will stay at coordinate one in X. It will use the same feed rate and linear movement as the previous line. **Y1.** will move to coordinate two point inches in Y. It will also use the same feed rate and linear movement as the previous line. ; End of block.

(N130 X1. Y0 ;)- N130 is the line number.**X1.** it will move to coordinate two point inches in X using the same feed and movement. **Y0** It will stay at coordinate two point inches in Y. ; End of block.

(N140 X0 Y-1. ;)- N140 is the line number. **X0** it will stay at coordinate two point inches in X. **Y-1.** It will move to coordinate one point inch in Y using the same feed and movement. ; End of block.

(N150 X-1. Y0 ;)- N150 is the line number. **X-1.** it will move to coordinate one point inch in X using the same feed and movement. **Y0** it will stay at coordinate one point inch in Y. ; End of block.

(**N160 G0 G28 Z0 M5 ;)- N160** is the line number. **G0** is rapid traverse. **G28** machine home coordinates. **Z0** is going to move to the Z zero coordinate in the machine home position. **M5** will turn off the spindle. ; End of block.

(**N170 X0 Y0 M9 ;)- N170** is the line number. **X0** is going to move to machine home zero position in rapid traverse. **Y0** is going to move to machine home zero position in rapid traverse. **M9** is going to turn off the coolant. ; End of block.

(**N180 M30 ;)- N180** is the line number. **M30** end of program and then reset itself at beginning of the program . ; End of block.

(**%)- %** This symbol is given at the end of every program to let the machine know that this is the end of the program. This symbol must be at the end of every program.

G41 cover compensation left and G42 cutter compensation right

In these two programs I'm going to show you how to use cutter compensation. These are very useful when you want to cut the profile of the part without having to do a large amount of calculations. If you need further information on which compensation to use and when please see page.

G41 cover compensation left

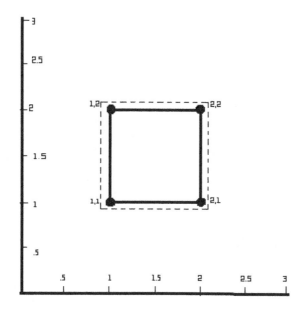

N10 O0002

N20 (Make sure tool two is 1/2 inch end mill);

N30 (Milling profile with half inch end mill);

N40 G91 G40 G80 G20;

N50 G0 G28 Z0;

N60 X0 Y0;

N70 T2;

N80 M6;

N90 G54 G41 D2 X1. Y.5 S1500 M3;

N100 G43 H1 Z.100 M8;

N110 G1 Z-.175 F80.;

N120 X0 Y1.5;

N140 X1. Y0;

N150 X0 Y-1.;

N160 X-1.5 Y0;

N170 G0 G28 Z0 M5;

N170 X0 Y0 M9;

N180 M2;

%

(N10 O0002)- N10 line number. **O0002** Program number.

(N20 (Make sure tool two is 1/2 inch end mill) ;)- **N20** line number. **(Make sure tool two is 1/2 inch end mill)** A note. ; End of block.

(N30 (Milling profile with half inch end mill) ;)- **N30** line number. **(Milling profile with half inch end mill)** A note. ; End of block.

(N40 G91 G40 G80 G20;)- **N40** line number. **G91** puts it in incremental mode. **G40** cancels cutter compensation. **G80** cancels canned cycle. **G20** put it in an inch format. ; End of block.

(N50 G0 G28 Z0 ;)- **N50** line number. **G0** is rapid traverse. **G28** use machine home zero coordinates. **Z0** Go to Z zero. ; End the block.

(N60 X0 Y0 ;)- **N60** line number. **X0** go to X zero. **Y0** go to Y zero. ; End of block.

(N70 T2 ;)- **N70** line number. **T2** call up tool two in the machine magazine. ; End of block.

(N80 M6 ;)- **N80** line number. **M6** preforms the tool change. ; End of block.

(N90 G54 G41 D2 X1. Y.5 S1500 M3 ;)- **N90** line number. **G54** use part home zero coordinates. G41 use cutter compensation left. **D2** used tool diameter offsets for tool two. **X1.** go to X one. **Y.5** go to Y point five. (Note) when preforming a milling operation it is better to start off the part then move into it at a set feed.

S1500 is spindle speed. **M3** turns on spindle. **;** End of block.

(N100 G43 H2 Z.100 M8 ;)- N100 line number. **G43** use tool offsets. **H2** used tool height offset for tool two. **Z.100** moves to point hundred above the part. **M8** turn on coolant. **;** End of block.

(N110 G1 Z-.175 F80. ;)- N110 line number. **G1** is a linear movement. **Z-.175** move negative point zero seventy-five into the part. **F80.** feed rate eighty. **;** End of block.

(N120 X0 Y1.5 ;)- N120 line number. **X0** stay at X one. **Y1.5** move to Y two. **;** End of block.

(N130 X1. Y0 ;)- N130 line number. **X1.** move to X two. **Y0** stay at Y two. **;** End of block.

(N140 X0 Y-1. ;)- N140 line number. **X0** stay at X two. **Y-1.** move to Y one. **;** End of block.

(N150 X-1.5 Y0 ;)- N150 line number. **X-1.5** move to X point five. (Note) when preforming a milling operation is better to move the tool all the way off the part when finished. **Y0** stay at Y one. **;** End of block.

(N160 G0 G40 G28 Z0 M5 ;)- N160 line number. **G0** is rapid traverse. **G40** is cancels cutter compensation. **G28** use machine home zero position coordinates. **Z0** move to Z zero. **M5** turn off spindle. **;** End of block.

(N170 X0 Y0 M9 ;)- N170 line number. **X0** move to X zero. **Y0** move to Y zero. **M9** turn coolant off. ; End of block.

(N180 M2 ;)- N180 line number. **M2** end of program. ; End of block.

(%)- % End of program.

G42 cutter compensation right

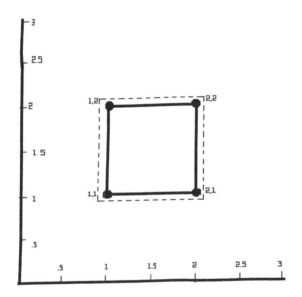

N10 O0003

N20 (Make sure tool two is 1/2 inch end mill);

N30 (Milling profile with half inch end mill);

N40 G91 G40 G80 G20;

N50 G0 G28 Z0;

N60 X0 Y0;

N70 T2;

N80 M6;

N90 G54 G42 D2 X.5 Y1. S1500 M3;

N100 G43 H2 Z.100 M8;

N110 G1 Z-.175 F80.;

N120 X1.5 Y0;

N130 X0 Y1.;

N140 X-1. Y0;

N150 X0 Y-1.5;

N160 G0 G40 G28 Z0 M5;

N170 X0 Y0 M9;

M180 M2;

%

(N10 O0003)- N10 line number. **O0003** Program number.

(N20 (Make sure tool two is 1/2 inch end mill) ;)- N20 line number. (Make sure tool two is 1/2 inch end mill) A note. ; End of block.

(N30 (Milling profile with half inch end mill) ;)- N30 line number. (Milling profile with half inch end mill) A note. ; End of block.

 (N40 G91 G40 G80 G20;)- N40 line number.G91 puts it in incremental mode.G40 cancels cutter compensation. G80 cancels canned cycle. G20 puts it in inch format. ; End of block.

(N50 G0 G28 Z0 ;)- N50 line number.G0 is rapid traverse. G28 use machine home zero position coordinates. Z0 Go to Z zero. ; End the block.

(N60 X0 Y0 ;)- N60 line number. X0 go to X zero.Y0 go to Y zero. ; End of block.

(N70 T2 ;)- N70 line number. T2 call up tool two in the machine magazine. ; End of block.

(N80 M6 ;)- N80 line number. M6 Preform a tool change. ; End of block.

 (N90 G54 G42 D2 X.5 Y1. S1500 M3 ;)- N90 line number. G54 use part home zero position coordinates. G42 use cutter compensation right. D2 used tool diameter offsets for tool two. X.5 go to X point five. (Note) when preforming a milling operation it is better to start off the part then feed into it. **Y1.** go to Y one. **S1500** spindle speed. **M3** turn on spindle. ; End of block.

(N100 G43 H2 Z.100 M8 ;)- N100 line number.**G43** tool offsets. **H2** used tool height offset for tool two. **Z.100** moves to point hundred above the part. **M8** turn on corporate. ; End of block.

(N110 G1 Z-.175 F80. ;)- N110 line number. **G1** is a linear movement. **Z-.175** move negative point zero seventy-five into the part. **F80.** set feed rate at eighty. ; End of block.

(N120 X1.5 Y0 ;)- N120 line number. **X1.5** move to X two. **Y0** stay at Y one. ; End of block.

(N130 X0 Y1. ;)- N130 line number. **X0** stay at X two. **Y1.** move to Y two. ; End of block.

(N150 X-1. Y0 ;)- N140 line number. **X-1.** move to X one. **Y0** stay at Y two. ; End of block.

(N150 X0 Y-1.5 ;)- N150 line number. **X0** stay at X one. **Y-1.5** move to Y point five. (Note) when preforming a milling operation is better to move the tool all the way off the part when finished. ; End of block.

(N160 G0 G40 G28 Z0 M5 ;)- N160 line number. **G0** is rapid traverse. **G40** is cancels cutter compensation. **G28** use machine home zero position coordinates. **Z0** move to Z zero. **M5** turn off spindle. ; End of block.

(N170 X0 Y0 M9 ;)- N170 line number. **X0** move to coordinates X zero. **Y0** move to coordinates Y zero. **M9** turn coolant off. ; End of block.

(M180 M2 ;)- N180 line number. **M2** end of program. ; End of block.

(%)- % End of the program.

G02 and G03 with R

G02 and G03 are radial movements which means they will move from one point to another and form of radius as they go. There are two ways of doing this using R along with the radius of the arc you wish to create is the simpler of the two ways. However it cannot be used for all circles or arts only those that are 180° or under. If you need more information please see page.

G02 with R

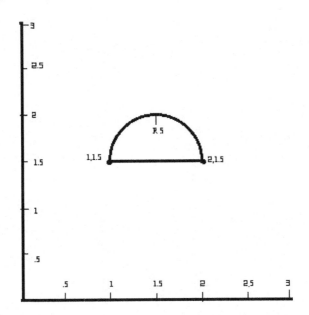

N10 O0004

N20 (Make sure tool one is a half inch spot drill);

N30 (Engraving with 1/2 inch spot drill);

N40 G91 G40 G80 G20;

N50 G0 G28 Z0;

N60 X0 Y0;

N70 T1;

N80 M6;

N90 G54 X1. Y1.5 S1500 M3;

N100 G43 H1 Z.100 M8;

N110 G1 Z-.175 F80.;

N120 G2 X1. Y0 R.5;

N130 G1 X-1. Y0;

N140 G0 G28 Z0 M9;

N150 X0 Y0 M5;

N160 M30;

%

(N10 O0004)- **N10** line number. **O0004** Program number.

(N20 (Make sure tool two is 1/2 inch spot drill) ;)- N20 line number. **(Make sure tool two is 1/2 inch spot drill)** A note. **;** End of block.

(N30 (Milling profile with half inch spot drill) ;)- N30 line number. (Milling profile with half inch spot drill) A note. ; End of block.

(N40 G91 G40 G80 G20 ;)- N40 line number. G91 puts it in incremental mode. G40 is cancels cutter compensation. G80 is cancels canned cycle. G20 puts it in inch format. ; End of block.

(N50 G0 G28 Z0 ;)- N50 line number. G0 is rapid traverse. G28 use machine home position coordinates. Z0 move to Z zero. ; End of block.

(N60 X0 Y0 ;)- N60 line number. X0 move to X zero. Y0 move to Y zero. ; End of block.

(N70 T1 ;)- N70 line number. T1 call up tool one in magazine. ; End of block.

(N80 M6 ;)- N80 line number. M6 preform tool change. ; End of block.

(N90 G54 X1. Y1.5 S1500 M3 ;)- N90 line number. G54 use part home position coordinates. X1. move to X one. Y1.5 move to Y one point five. S1500 spindle speed. M3 turn on spindle. ; End of block.

(N100 G43 H1 Z.100 M8 ;)- N100 line number. G43 used tool length offset. H1 use height offset for tool one. Z.100 move to point one hundred above the part. M8 turn on coolant. ; End of block.

(N110 G1 Z-.175 F80. ;)- N110 line number. **G1** is linear movement. **Z-.175** move to negative point zero seventy-five into the part. **F80.** set feed rate. **;** End of block.

(N120 G2 X1. Y0 R.5 ;)- N120 line number. **G2** this is a movement that will give a clockwise radius. (Note) G2 is a modal code and will remain active unless given another code to cancel it. If the feed rate is specified earlier in the program with a G01 the feed rate will remain the same unless otherwise specified when you use a G02 or G03. **X1.** move to X two. **Y0** stay at Y one point five. **R.5** the arc has a radius of point five. **;** End of block.

(N130 G1 X-1. Y0 ;)- N130 line number. **G1** is a linear movement. **X-1.** move to X one. **Y0** stay at Y one point five. **;** End of block.

(N140 G0 G28 Z0 M5 ;)- N140 line number. **G0** is rapid traverse.**G28** use machine home position coordinates. **Z0** move to Z zero. **M5** turn off spindle. **;** End of block.

(N150 X0 Y0 M9 ;)- N150 line number. **X0** move to X zero. **Y0** move to Y zero. **M9** turn off the coolant. **;** End of block.

(N160 M30 ;)- N160 line number. **M30** end of program and reset at beginning. **;** End of block.

(%)- % End of the program.

G03 with R

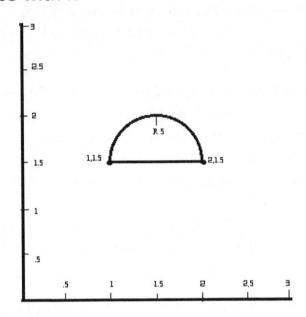

N10 O0005

N20 (Make sure tool one is a half inch spot drill);

N30 (Engraving with 1/2 inch spot drill);

N40 G91 G40 G80 G20;

N50 G0 G28 Z0;

N60 X0 Y0;

N70 T1;

N80 M6;

N90 G0 G54 X1. Y1.5 S1500 M3;

N100 G43 H1 Z.100 M8;

N110 G1 Z-.175 F80.;

N120 G1 X1. Y0;

N130 G3 X-1. Y0 R.5;

N140 G0 G28 Z0 M9;

N150 X0 Y0 M5;

N160 M30;

%

(N10 O0005)- N10 line number. **O0005** Program number.

(N20 (Make sure tool two is 1/2 inch spot drill) ;)- N20 line number. **(Make sure tool two is 1/2 inch spot drill)** A note. **;** End of block.

(N30 (Milling profile with half inch spot drill) ;)- N30 line number. **(Milling profile with half inch spot drill)** A note. **;** End of block.

(N40 G91 G40 G80 G20;)- N40 line number. **G91** puts it in incremental mode. G40 is cancels cutter compensation. **G80** is cancels canned cycle. **G20** puts it in inch format. **;** End of block.

(N50 G0 G28 Z0 ;)- N50 line number. **G0** is rapid traverse. **G28** use machine home position coordinates. **Z0** move to Z zero. ; End of block.

(N60 X0 Y0 ;)- N60 line number. **X0** move to X zero. **Y0** move to Y zero. ; End of block.

(N70 T1 ;)- N70 line number. **T1** call up tool one in magazine. ; End of block.

(N80 M6 ;)- N80 line number. **M6** preform tool change. ; End of block.

(N90 G54 X1. Y1.5 S1500 M3 ;)- N90 line number. **G54** use part home position coordinates. **X1.** move to X one.**Y1.5** move to Y one point five. **S1500** spindle speed. **M3** turn on spindle. ; End of block.

(N100 G43 H1 Z.100 M8 ;)- N100 line number. **G43** used tool length offset. **H1** use height offset for tool one. **Z.100** move to point one hundred above the part. **M8** turn on coolant. ; End of block.

(N110 G1 Z-.175 F80. ;)- N110 line number. **G1** is linear movement. **Z-.175** move to negative point zero seventy-five into the part. **F80.** set feed rate. ; End of block.

(N120 G1 X1. Y0 ;)- N120 line number. **G1** is a linear movement. **X1.** move to X two. **Y0** stay at Y one point five. ; End of block.

(N130 G3 X-1. Y0 R.5 ;)- **N130** line number. **G3** this is a movement that will give a counter clockwise radius. (Note) G3 is a modal code and will remain active unless given another code to cancel it. If the feed rate is specified earlier in the program with a G01 the feed rate will remain the same unless otherwise specified when you use a G02 or G03. **X1.** move to X one. **Y0** stay at Y one point five. **R.5** the arc has a radius of point five. ; End of block.

(N140 G0 G28 Z0 M9 ;)- **N140** line number. **G0** is rapid traverse. **G28** use machine home position coordinates. **Z0** move to Z zero. **M9** turn off the coolant. ; End of block.

(N150 X0 Y0 M5 ;)- **N150** line number. **X0** move to X zero. **Y0** move to Y zero. M5 turn off spindle. ; End of block.

(N160 M30 ;)- **N160** line number. **M30** end of program and reset at beginning. ; End of block.

(%)- % End of the program.

G02 and G03 with I and J

Using I and J is the second method of using G02
or G03. It can be used on any degree of arc or circle but
does take a little more work to use If you need more
information please see page.

G02 with I and J

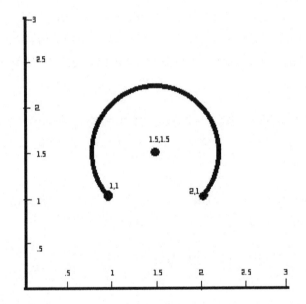

N10 O0006

N20 (Make sure tool one is a half inch spot drill);

N30 (Engraving with 1/2 inch spot drill);

N40 G91 G40 G80 G20;

N50 G0 G28 Z0;

N60 X0 Y0;

N70 T1;

N80 M6;

N90 G0 G54 X1. Y1. S1500 M3;

N100 G43 H1 Z.100 M8;

N110 G1 Z-.175 F80.;

N120 G02 X1. Y0 I .5 J.5;

N130 G0 G28 Z0 M9;

N140 X0 Y0 M5;

N150 M2;

%

(**N10 O0006**)- **N10** line number. **O0006** Program number.

(**N20 (Make sure tool two is 1/2 inch spot drill) ;**)- **N20** line number. (**Make sure tool two is 1/2 inch spot drill**) A note. ; End of block.

(**N30 (Milling profile with half inch spot drill) ;**)- **N30** line number. (**Milling profile with half inch spot drill**) A note. ; End of block.

(N40 G91 G40 G80 G20 ;)- **N40** line number. **G91** puts it in incremental mode. **G40** is cancels cutter compensation. **G80** is cancels canned cycle. **G20** puts it in inch format. ; End of block.

(N50 G0 G28 Z0 ;)- **N50** line number. **G0** is rapid traverse. **G28** use machine home position coordinates. **Z0** move to Z zero. ; End of block.

(N60 X0 Y0 ;)- **N60** line number.**X0** move to X zero. **Y0** move to Y zero. ; End of block.

(N70 T1 ;)- **N70** line number. **T1** call up tool one in magazine. ; End of block.

(N80 M6 ;)- **N80** line number. **M6** preform tool change. ; End of block.

(N90 G54 X1. Y1. S1500 M3 ;)- **N90** line number. **G54** use part home position coordinates. **X1.** move to X one.**Y1.** move to Y one. **S1500** spindle speed. **M3** turn on spindle. ; End of block.

(N100 G43 H1 Z.100 M8 ;)- **N100** line number. **G43** used tool length offset. **H1** use height offset for tool one. **Z.100** move to point one hundred above the part. **M8** turn on coolant. ; End of block.

(N110 G1 Z-.175 F80. ;)- **N110** line number. **G1** is linear movement. **Z-.175** move to negative point zero seventy-five into the part. **F80.** set feed rate. ; End of block.

(N120 G2 X1. Y0 I.5 J.5 ;)- **N120** line number. **G2** it will move in a clockwise direction. **X1.** go to X two. **Y0** stay at Y one. **I.5** I represents the distance between the starting point of the arc to the center of the arc in X. Distance is always calculated in an incremental format even in absolute mode. **J.5** J represents the distance between the starting point of the arc to the center of the arc in Y. The distance is always calculated in an incremental format even in absolute mode. ; End of block.

(N130 G0 G28 Z0 M9 ;)- **N130** line number. **G0** is rapid traverse. **G28** use machine home position coordinates. **Z0** move to Z zero. **M9** turn off the coolant. ; End of block.

(N140 X0 Y0 M5 ;)- **N140** line number. **X0** move to X zero. **Y0** move to Y zero. **M5** turn off spindle. ; End of block.

(N150 M2 ;)- **N150** line number. **M2** end of program. ; End of block.

(%)- % End of the program.

G03 with I and J

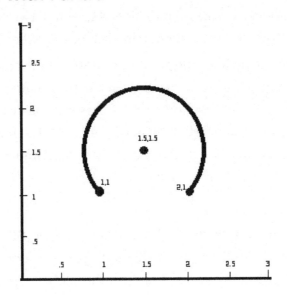

N10 O0007

N20 (Make sure tool one is a half inch spot drill);

N30 (Engraving with 1/2 inch spot drill);

N40 G91 G40 G80 G20;

N50 G0 G28 Z0;

N60 X0 Y0;

N70 T1;

N80 M6;

N90 G0 G54 X2. Y1. S1500 M3;

N100 G43 H1 Z.100 M8;

N110 G1 Z-.175 F80.;

N120 G03 X-1. Y0 I-.5 J.5;

N130 G0 G28 Z0 M9;

N140 X0 Y0 M5;

N150 M2;

%

(N10 O0007)- **N10** line number. **O0007** Program number.

(N20 (Make sure tool two is 1/2 inch spot drill) ;)- **N20** line number. **(Make sure tool two is 1/2 inch spot drill)** A note. ; End of block.

(N30 (Milling profile with half inch spot drill) ;)- **N30** line number. **(Milling profile with half inch spot drill)** A note. ; End of block.

(N40 G91 G40 G80 G20 ;)- **N40** line number. **G91** puts it in incremental mode. **G40** is cancels cutter compensation. **G80** is cancels canned cycle. **G20** puts it in inch format. ; End of block.

(N50 G0 G28 Z0 ;)- **N50** line number. **G0** is rapid traverse. **G28** use machine home position coordinates. **Z0** move to Z zero. ; End of block.

(N60 X0 Y0 ;)- **N60** line number. **X0** move to X zero. **Y0** move to Y zero. ; End of block.

(N70 T1 ;)- N70 line number. **T1** call up tool one in magazine. ; End of block.

(N80 M6 ;)- N80 line number. **M6** preform tool change. ; End of block.

(N90 G54 X2. Y1. S1500 M3 ;)- N90 line number. **G54** use part home position coordinates. **X2.** move to X two.**Y1.** move to Y one. **S1500** spindle speed. **M3** turn on spindle. ; End of block.

(N100 G43 H1 Z.100 M8 ;)- N100 line number. **G43** used tool length offset. **H1** use height offset for tool one. **Z.100** moves to point one hundred above the part. **M8** turn on coolant. ; End of block.

(N110 G1 Z-.175 F80. ;)- N110 line number. **G1** is linear movement. **Z-.175** move to negative point zero seventy-five into the part. **F80.** set feed rate. ; End of block.

(N120 G3 X-1. Y0 I -.5 J.5 ;)- N120 line number. **G3** it will move in a counter clockwise direction. **X-1** go to X one. **Y0** stay at Y one. **I-.5** I represents the distance between the starting point of the arc to the center of the arc in X. The distance is always calculated in an incremental format even in absolute mode. **J.5** J represents the distance between the starting point of the arc to the center of the arc in Y. The distance is always calculated in an incremental format even in absolute mode. ; End of block.

(N130 G0 G28 Z0 M9 ;)- **N130** line number. **G0** is rapid traverse.**G28** use machine home position coordinates. **Z0** move to Z zero. **M9** turn off the coolant. ; End of block.

(N140 X0 Y0 M5 ;)- **N140** line number. **X0** move to X zero. **Y0** move to Y zero. **M5** turn off spindle. ; End of block.

(N150 M2 ;)- **N150** line number. **M2** end of program. ; End of block.

(%)- **%** End of the program.

Using canned drill cycles

Using a canned drilling cycles makes drilling a large number of holes quite easy since you do not have to reenter the information about the hole at each point. You simply enter the information once then simply give it the next location you wish to the drill.

Canned drilling cycles

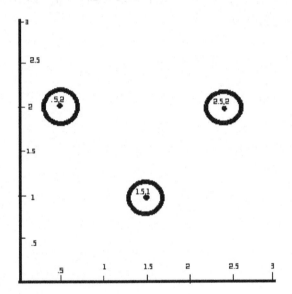

N10 O0008;

N20 (Tool 1 needs to be 1/2 inch spot drill);

N30 (Tool 3 needs to be a 27/64 inch drill);

N40 (Tool 1 Spot Drill 1/2 inch);

N50 G91 G40 G80 G20;

N60 G0 G28 Z0;

N70 X0 Y0;

N80 T1;

N90 M6;

N100 G54 X.5 Y2. S1500 M3;

N110 G43 H1 Z.100 M8 G4 P500;

N120 G81 G99 Z-.125 R.125 F80.;

N130 X1. Y-1.;

N140 X1. Y1.;

N150 G80;

N160 G0 G28 Z0 M5;

N170 X0 Y0 M9;

N180 M1;

N190 (Tool 3 Drill 27/64);

N200 G91 G40 G80 G20;

N210 G0 G28 Z0;

N220 X0 Y0;

N230 T3;

N240 M6;

N250 G54 X.5 Y2. S1500 M3;

N260 G43 H3 Z.100 M8 G4 P500;

N270 G83 G99 Z-.600 Q.100 R.600 F80.;

N280 /X1. Y-1.;

N290 /X1. Y1.;

N300 G80;

N310 G0 G28 Z0 M5;

N320 X0 Y0 M9;

N330 M1;

N340 (Tool 1 Chamfer 1/2 Spot Drill);

N350 G91 G40 G80 G20;

N360 G0 G28 Z0;

N370 X0 Y0;

N380 T1;

N390 M6;

N400 G54 X.5 Y2. S1500 M3;

N410 G43 H1 Z.100 M8 G4 P500;

N420 G81 G99 Z-.175 R.175 F80.;

N430 X1. Y-1.;

N440 X1. Y1.;

N450 G80;

N460 G0 G28 Z0 M5;

N470 X0 Y0 M9;

N480 M30;

%

(N10 O0008 ;)- N10 line number. O0008 Program number.

(N20 (Tool 1 need to be 1/2 inch spot drill) ;)- N20 line number. (Tool 1 needs to be 1/2 inch spot drill) A note. ; End the block.

(N30 (Tool 3 needs to be a 27/64 inch drill) ;) – N30 line number. (Tool 3 needs to be a 27/64 inch drill) A note. ; End of block.

(N40 (Spot Drill 1/2 inch) ;)- N40 line number. (Spot Drill 1/2 inch) A note. ; End of block.

(N50 G91 G40 G80 G20 ;)- N50 line number. G91 puts it in incremental mode. G40 is cancels cutter compensation. G80 is cancels canned cycle. G20 puts it in inch format. ; End of block.

(N60 G0 G28 Z0 ;)- N60 line number. G0 is rapid traverse. G28 use machine home position coordinates. Z0 move to Z zero. ; End of block.

(N70 X0 Y0 ;)- **N70** line number.**X0** move to X zero. **Y0** move to Y zero. **;** End of block.

(N80 T1 ;)- **N80** line number. **T1** call up tool one in magazine. **;** End of block.

(N90 M6 ;)- **N90** line number. **M6** preform tool change. **;** End of block.

(N100 G54 X.5 Y2. S1500 M3 ;)- **N100** line number. **G54** use part home position coordinates. **X.5** move to X point five. **Y2.** move to Y two. **S1500** spindle speed. **M3** turn on spindle. **;** End of block.

(N110 G43 H1 Z.100 M8 G4 P500 ;)- **N110** line number. **G43** used tool length offset. **H1** use height offset for tool one. **Z.100** move to point one hundred above the part. **M8** turn on coolant.**G4** dwell. **P500** the amount of time for dwell. **;** End of block.

(N120 G81 G99 Z-.125 R.100 F80. ;)- **N120** line number.**G81** is a canned drill cycle. This is a modal code, and will perform the same drilling action that is specified on this line at each location without need of further input on the other lines. **G99** Is R level return. This tells the machine to use R for the return height. **Z-.125** will move negative point zero twenty-five into the part. **R.100** the R in this case stands for retract height. Which means the height at which it will return after it is done drilling the hole. Which in this case is .100 above the part (when using G99). **F80.** set feed rate at eighty. **;** End the block.

(N130 X1. Y-1. ;)- N130 line number. **X1.** move to X one point five. **Y-1.** move to Y one. **;** End of block.

(N140 X1. Y1. ;)- N140 line number. **X1.** move to X two point five. **Y1.** move to Y two. **;** End of block.

(N150 G80 ;)- N150 line number. **G80** cancels canned cycle. **;** End of block.

(N160 G0 G28 Z0 M5 ;)- N160 line number. **G0** is rapid traverse. **G28** use machine home position coordinates. **Z0** move to Z zero. **M5** turn off spindle. **;** End of block.

(N170 X0 Y0 M9 ;)- N170 line number. **X0** move to X zero. **Y0** move to Y zero. **M9** turn off the coolant. **;** End of block.

(N180 M1 ;)- N180 line number. **M1** is optional stop. This will allow you to stop the machine at this point as long as the optional stop button is pressed on the machine. It is good to place an M1 at the end of operations that way you can stop the machine and check the part if needed. It also gives you a way to stop the machine without having to wait for the completion of the program. **;** End the block.

(N190 (Drill 27/64) ;)- N190 line number. **(Drill 27/64)** A note. **;** End the block.

(N200 G91 G40 G80 G20 ;)- N200 line number. **G91** puts it in incremental mode. **G40** is cancels cutter compensation. **G80** is cancels canned cycle. **G20** puts it in inch format. ; End of block.

(N210 G0 G28 Z0 ;)- **N210** line number. **G0** is rapid traverse. **G28** use machine home position coordinates. **Z0** move to Z zero. ; End of block.

(N220 X0 Y0 ;)- **N220** line number. **X0** move to X zero. **Y0** move to Y zero. ; End of block.

(N230 T1 ;)- **N230** line number. **T1** call up tool one in magazine. ; End of block.

(N240 M6 ;)- **N240** line number. **M6** preform tool change. ; End of block.

(N250 G54 X.5 Y2. S1500 M3 ;)- **N250** line number. **G54** use part home position coordinates. **X.5** move to X point five. **Y2.** move to Y two. **S1500** spindle speed. **M3** turn on spindle. ; End of block.

(N260 G43 H1 Z.100 M8 G4 P500 ;)- **N260** line number. **G43** used tool length offset. **H1** use height offset for tool one. **Z.100** moves to point one hundred above the part. **M8** turn on coolant. **G4** dwell. **P500** the amount of time for dwell. ; End of block.

(N270 G83 G98 Z-.600 Q.100 R.100 F80. ;)- **N270** line number. **G83** is a canned drill cycle. This is a modal code, and will perform the same drilling action that is specified on this line at each location without need of further input on the other lines. This code is known as a peck drill cycle. It will travel so far into the part before backing off to break the chip then continuing to drill till it reaches the next target increment or finishes the hole. **G98** is initial level return. It tell the machine to use the last Z height instead of the R return height. **Z-.600** will move negative point zero five hundred into the part. **Q.100** Q is how you set the target increment for the Peck drilling cycle in this case is set to .100 which mean it will perform a chip break every hundred thousandths into the part. **R.100** return height point one hundred (when using a G99) but is not used in the operation. **F80.** set feed rate at eighty. ; End the block.

(N280 /X1. Y-1. ;)- **N280** line number. **/** When the block delete button is activated this will tell the machine not to read this line of programming. This is useful when you wish to only run one of a particular action so you can check accuracy or quality of the process. When the block delete button is not on the machine will read this line of programming normally. **X1** moved to X one point five. **Y-1** move to Y one. ; End of block.

(N290/ X1. Y1. ;)- **N290** line number. **/**Skip this line of programming when the block delete button is active. **X1.** move to X two point five. **Y1.** move to Y two. ; End of block.

(N300 G80 ;)- **N300** line number. **G80** cancels canned cycle. ; End of block.

(N310 G0 G28 Z0 M5 ;)- **N310** line number. **G0** is rapid traverse. **G28** use machine home position coordinates. **Z0** moved to Z zero. **M5** turn off spindle. ; End of block.

(N320 X0 Y0 M9 ;)- **N320** line number. **X0** move to X zero. **Y0** move to Y zero. **M9** turn off the coolant. ; End of block.

(N330 M1 ;)- **N330** line number. **M1** is optional stop. ; End of block.

(N340 (Chamfer 1/2 Spot Drill) ;)- **N340** line number. **(Chamfer 1/2 Spot Drill)** A note. ; End of block.

(N350 G91 G40 G80 G20 ;)- **N350** line number. **G91** puts it in incremental mode. **G40** is cancels cutter compensation. **G80** is cancels canned cycle. **G20** puts it in inch format. ; End of block.

(N360 G0 G28 Z0 ;)- **N360** line number. **G0** is rapid traverse. **G28** use machine home position coordinates. **Z0** move to Z zero. ; End of block.

(N370 X0 Y0 ;)- **N370** line number. **X0** move to X zero. **Y0** move to Y zero. ; End of block.

(N380 T1 ;)- **N380** line number. **T1** call up tool one in magazine. ; End of block.

(N390 M6 ;)- **N390** line number. **M6** preform tool change. ; End of block.

(N400 G54 X.5 Y2. S1500 M3 ;)- N400 line number. **G54** use part home position coordinates. **X.5** move to X point five. **Y2.** move to Y two. **S1500** spindle speed. **M3** turn on spindle. ; End of block.

(N410 G43 H1 Z.100 M8 G4 P500 ;)- N410 line number. **G43** used tool length offset. **H1** use height offset for tool one. **Z.100** moves to point one hundred above the part. **M8** turn on coolant.**G4** dwell. **P500** the amount of time for dwell. ; End of block.

(N420 G81 G99 Z-.175 R.100 F80. ;)- N420 line number. **G81** is a canned drill cycle. **G99** Is R level return. **Z-.175** will move negative point zero seventy-five into the part. **R.100** return height to point one hundred. **F80.** set feed rate at eighty. ; End of block.

(N430 X1. Y-1. ;)- N430 line number. **X1.** move to X one point five. **Y-1.** move to Y one. ; End of block.

(N440 X1. Y1. ;)- N440 line number. **X1.** move to X two point five. **Y1.** move to Y two. ; End of block.

(N450 G80 ;)- N450 line number. **G80** cancels canned cycle. ; End of block.

(N460 G0 G28 Z0 M5 ;)- N460 line number. **G0** is rapid traverse. **G28** use machine home position coordinates. **Z0** move to Z zero. **M5** turn off spindle. ; End of block.

(N470 X0 Y0 M9 ;)- **N470** line number. **X0** move to X zero. **Y0** move to Y zero. **M9** turn off the coolant. ; End of block.

(N480 M30 ;)- N480 line number. **M30** end of program and reset at beginning. ; End of block.

(%)- % Symbol for in the program.

Scenarios

Exercises

Now you are familiar with how to write a program, here are a couple of scenarios so you can write a couple programs on your own. I will give you the scenario and the picture of the tool path then you can write program based on it. After that you can turn to appendix A and compare them to mine.

Milling operation

In this operation you are going to be milling down a one inch block. You are going to take off .050. You will be using two tools. Tool 2 will be a half inch rough mill and you are going to take two passes with it. The first one is going to be .015 into the part and the second one will be .025 into the part. Next you will swap out tools for tool 4 which is going to be a half inch finish mill. And you will take two .005 passes with it.

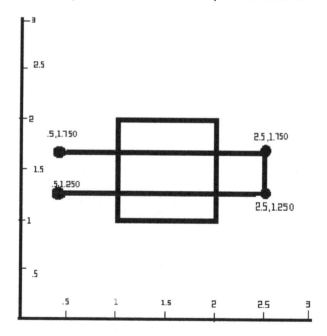

All smiles

In this operation you will be engraving this happy face with tool 1 which is a half inch spot drill. You will be going into the part .075. You will start by engraving the outside circle. Then you will spot drill the eyes and nose. And the last thing you will do is the mouth.

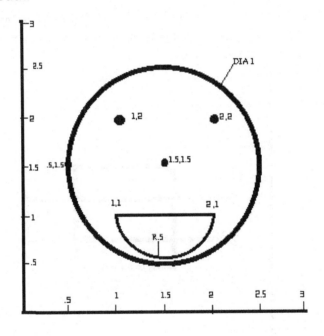

Conclusion

In conclusion I hope this book was a great help to you in understand the world of CNC programming. And I want to finish by saying good luck in your future projects.

Appendix A

Milling operation (absolute)

O0010

(Make sure tool 2 is 1/2" roughing mill);

(Make sure tool 4 is 1/2" finish mill);

(1/2 " Rough mill)

N10 G90 G40 G80 G20;

N20 G0 G28 Z0;

N30 X0 Y0;

N40 T2;

N50 M6;

N60 G54 X.5 Y1.75 S1500 M3;

N70 G43 H2 Z.100 M8;

N80 G1 Z-.015 F80.;

N90 X2.5 Y1.75;

N100 X2.5 Y1.25;

N110 X.5 Y1.25;

N120 X.5 Y1.75;

N130 Z-.040;

N140 X2.5 Y1.75;

N150 X2.5 Y1.25;

N160 X.5 Y1.25;

N170 G0 G28 Z0 M5;

N180 X0 Y0 M9;

(1/2" Finish mill)

N190 G90 G40 G80 G20;

N200 G0 G28 Z0;

N210 X0 Y0;

N220 T4;

N230 M6;

N240 G54 X.5 Y1.75 S1600 M3;

N250 G43 H4 Z.100 M8;

N260 G1 Z-.045 F60.;

N270 X2.5 Y1.75;

N280 X2.5 Y1.25;

N290 X.5 Y1.25;

N300 X.5 Y1.75;

N310 Z-.050;

N320 X2.5 Y1.75;

N330 X2.5 Y1.25;

N340 X.5 Y1.25;

N350 G0 G28 Z0 M5;

N360 X0 Y0 M9;

N370 M2;

%

Milling operation (incremental)

O0010

(Make sure tool 2 is 1/2" roughing mill);

(Make sure tool 4 is 1/2" finish mill);

(1/2 " Rough mill)

N10 G91 G40 G80 G20;

N20 G0 G28 Z0;

N30 X0 Y0;

N40 T2;

N50 M6;

N60 G54 X.5 Y1.75 S1500 M3;

N70 G43 H2 Z.100 M8;

N80 G1 Z-.115 F80.;

N90 X2. Y0;

N100 X0 Y-.5;

N110 X-2. Y0;

N120 X0 Y.5;

N130 Z-.025;

N140 X2. Y0;

N150 X0 Y-.5;

N160 X-2. Y0;

N170 G0 G28 Z0 M5;

N180 X0 Y0 M9;

(1/2" Finish mill)

N190 G91 G40 G80 G20;

N200 G0 G28 Z0;

N210 X0 Y0;

N220 T4;

N230 M6;

N240 G54 X.5 Y1.75 S1600 M3;

N250 G43 H4 Z.100 M8;

N260 G1 Z-.145 F60.;

N270 X2. Y0;

N280 X0 Y-.5;

N290 X-2. Y0;

N300 X0 Y.5;

N310 Z-.005;

N320 X2. Y0;

N330 X0 Y-.5;

N340 X-2. Y0;

N350 G0 G28 Z0 M5;

N360 X0 Y0 M9;

N370 M2;

%

All smiles (absolute)

O0011

(Make sure tool one is half inch spot drill)

(Engraving with half inch spot drill)

N10 G90 G40 G80 G20;

N20 G0 G28 Z0;

N30 X0 Y0 ;

N40 T1;

N50 M6;

N60 G54 G0 X.5 Y1.5 S1500 M3;

N70 G43 H1 Z.100 M8;

N80 G01 Z-.075 F80.;

N90 G02 X.5 Y1.5 I1 J0;

N100 G0 Z.1;

N110 X1. Y2.;

N120 G81 G98 Z-.075 R.100 F80.;

N130 X2. Y2.;

N140 X1.5 Y1.5;

N150 G80;

N160 G1 X1. Y1. F80.;

N170 Z-.075;

N180 X2. Y1.;

N190 G3 X1. Y1. R.5;

N200 G0 G28 Z0 M5;

N210 X0 Y0 M9;

N220 M2;

%

All smiles (incremental)

O0011

(Make sure tool one is half inch spot drill)

(Engraving with half inch spot drill)

N10 G91 G40 G80 G20;

N20 G0 G28 Z0;

N30 X0 Y0 ;

N40 T1;

N50 M6;

N60 G54 G0 X.5 Y1.5 S1500 M3;

N70 G43 H1 Z.100 M8;

N80 G01 Z-.175 F80.;

N90 G02 X.0 Y0 I1 J0;

N100 G0 Z.175;

N110 X.5 Y.5;

N120 G81 G98 Z-.175 R.175 F80.;

N130 X1. Y0;

N140 X-.5 Y-.5;

N150 G80;

N160 G1 X-.5 Y-.5 F80.;

N170 Z-.175;

N180 X1. Y0;

N190 G3 X-1. Y0 R.5;

N200 G0 G28 Z0 M5;

N210 X0 Y0 M9;

N220 M2;

%

Appendix B

Fraction to Dismals

8ths	16ths
1/8=.125	1/16=.062
1/4=.250	3/16=.187
3/8=.375	5/16=.312
1/2=.500	7/16=.437
5/8=.625	9/16=.562
3/4=.750	11/16=.687
7/8=.875	13/16=.812
	15/16=.937

32nds

1/32=.032

3/32=.093

5/32=.156

7/32=.218

9/32=.281

11/32=.343

13/32=.406

15/32=.468

17/32=.531

19/32=.593

21/32=.656

23/32=.718

25/32=.781

27/32=.843

29/32=.906

31/32=.968

64ths

1/64=.015

3/64=.046

5/64=.078

7/64=.109

9/64=.140

11/64=.171

13/64=.203

15/64=.234

17/64=.265

19/64=.296

21/64=.328

23/64=.359

25/64=.390

27/64=.421

29/64=.453

31/64=.484

33/64=.515

35/64=.546

37/64=.578

39/64=.609

41/64=.640

43/64=.671

45/64=.703

47/64=.734

49/64=.765

51/64=.796

53/64=.828

55/64=.859

57/64=.890

59/64=.921

61/64=.953

63/64=.984

Made in the USA
Coppell, TX
09 April 2021